I0494104

Disclaimer

The publisher of this book is by no way associated with the National Institute of Standards and Technology (NIST). The NIST did not publish this book. It was published by 50 page publications under the public domain license.

50 Page Publications.

Book Title: Factors Influencing the Smoldering Performance of Polyurethane Foam

Book Author: Mauro Zammarano; Szabolcs S. Matko; Roland H. Kraemer; Rick D. Davis; Jeffrey W. Gilman; Shivani N. Mehta;

Book Abstract: The objective of this study was to evaluate whether polyurethane foam (PUF) could be produced as a standard reference material for ultimate use in a standard test intended to ensure the smoldering performance of commercially available upholstered furniture. For this purpose, PUF was produced by a commercial manufacturer and its smoldering propensity was evaluated. The experimental design was organized into four parts or Iterations. In Iteration 1, the results showed that smoldering intensity was dominated by foam morphology, which overrides effects of chemical composition for the range of polyols and surfactants studied here. In Iteration 2, the morphology of the foam was controlled by varying the processing parameters, i.e., tin catalyst content, water content and mixing head pressure. The results showed that smoldering increased with air permeability but a better morphological descriptor of the foam structure was required to fully characterize smoldering in the high permeability range (i.e., in PUF with a dominantly open-cell structure). In Iteration 3, it was determined that for these types of foams, smoldering was controlled by cell size. The feasibility of a reference material with well-characterized and reproducible smoldering is linked to the ability to manufacture an open-cell PUF with a well-defined cell size and density. In Iteration 4, scale-up from the pilot plant to the production line was considered. Foams with much more consistent properties (e.g., air permeability, density, cell size, etc.) throughout the bun of foam were fabricated in the production line as compared to the pilot plant. This allowed for the fabrication of PUFs with a very reproducible smoldering to be used as a potential standard reference material.

Citation: NIST TN - 1747

Keywords: Polyurethane; foam; smoldering

NIST Technical Note 1747

Factors Influencing the Smoldering Performance of Polyurethane Foam

Mauro Zammarano
Roland H. Krämer
Szabolcs Matko
Shivani Mehta
Jeffrey W. Gilman
Rick D. Davis

http://dx.doi.org/10.6028/NIST.TN.1747

National Institute of Standards and Technology
U.S. Department of Commerce

NIST Technical Note 1747

Factors Influencing the Smoldering Performance of Polyurethane Foam

Mauro Zammarano
Roland H. Krämer[*]
Szabolcs Matko
Rick D. Davis
Fire Research Division
Engineering Laboratory
National Institute of Standards and Technology
**Former Guest Researcher*

Jeffrey W. Gilman
Polymers Division
Material Measurement Laboratory
National Institute of Standards and Technology

Shivani Mehta
Directorate for Engineering Sciences
U.S. Consumer Product Safety Commission

http://dx.doi.org/10.6028/NIST.TN.1747

July 2012

U.S. Department of Commerce
Rebecca Blank, Acting Secretary

National Institute of Standards and Technology
Patrick D. Gallagher, Under Secretary of Commerce for Standards and Technology and Director

National Institute of Standards and Technology Technical Note 1747
Natl. Inst. Stand. Technol. Tech. Note 1747, 70 pages (July 2012)
http://dx.doi.org/10.6028/NIST.TN.1747
CODEN: NTNOEF

Acknowledgements

Appreciation is extended to the Consumer Product Safety Commission for financial and technical support for this project under Interagency Agreement Number: CPSC-I-09-0015.

Abstract

The objective of this study was to evaluate whether polyurethane foam (PUF) could be produced as a Standard Reference Material for ultimate use in a standard test intended to ensure the smoldering performance of commercially available upholstered furniture. For this purpose, PUF was produced by a commercial manufacturer and its smoldering propensity was evaluated. The experimental design was organized into four parts or "Iterations". In Iteration 1, the results showed that smoldering intensity was dominated by foam morphology (cell size, strut thickness and length, open vs. closed cell structure, etc.), which overrode effects of chemical composition for the range of polyols and surfactants studied. In Iteration 2, the morphology of the foam was controlled by varying the processing parameters, that is, the tin catalyst content, water content and mixing head pressure. The results showed that smoldering increased with air permeability but that a better morphological descriptor of the foam structure was required to fully characterize smoldering in the high permeability range (*i.e.*, in PUF with a dominantly open-cell structure). In Iteration 3, it was determined that for these PUFs, smoldering was controlled by cell size. The feasibility of a reference material with well-characterized and reproducible smoldering is linked to the ability to manufacture an open-cell PUF with a well-defined cell size and density. Preliminary data from Iteration 4 indicates that PUFs with consistent smoldering can be produced on a commercial scale manufacturing line.

Table of Contents

Executive Summary

The objective of this project was to evaluate the potential for specifying and obtaining flexible polyurethane foam (PUF) that smoldered readily and consistently. Such a foam could then be used as a Standard Reference Material (SRM) for evaluating upholstery fabrics for their tendency to support smoldering and, thus, reduce the ignition potential of upholstered furniture by cigarettes.

Smoldering is a self-sustaining oxidation process through which heat is released from surface reaction of the fuel with oxygen in the air. The chemical reaction is slow compared to a flaming reaction, and the heat released is far lower. To maintain the fuel at a smoldering temperature the released heat must be retained inside the fuel. Much of the heat from any reaction at the outer surface of the fuel tends to be dissipated to the surroundings. The surrounding-fuel-mass acts as a thermal insulator. Once oxygen has reached the heated interior surfaces of the fuel-mass, it is completely consumed in the smoldering reaction. The oxygen must be replenished in order for the smoldering to continue. Thus, fuels with a tendency to smolder have pathways for air to penetrate to their interior. The heat generation rate then depends on the chemistry of the fuel, its porosity (the magnitude of the internal surface area), and the permeability of the fuel to air.

For this project, batches of PUF were prepared according to NIST specifications and were obtained from a commercial manufacturer ("foamer"). Each batch was characterized by combustion tests. In addition, the air permeability for each batch of foam was measured using the pressure drop across a slice of set thickness and area. The foam firmness was measured using a standard indentation force deflection device. The internal surface area per unit volume of the foam was calculated from the measured Brunauer-Emmett-Teller (BET) surface adsorption and the foam density. The cell size distribution was determined through microscopic analysis.

PUFs were designed for ultimate use as a reference material in the proposed CPSC standard 16 C.F.R. 1634 (herein referred to as "CPSC proposed standard"). The possible implications for open-flame configuration, which are part of the CPSC proposed standard, also needed to be considered. For this reason, even though smoldering propensity was the main focus of this study, open-flame combustion was also investigated and specimens from each of the batches were examined using four combustion methods:

- The method to be used in the CPSC proposed regulation for testing the contribution of an upholstery fabric to the smoldering of a fabric/PUF composite (herein referred to as "mockup test"). The orientation of the materials tested in this method replicates a common scenario for the ignition of upholstered furniture. Two pieces of PUF are placed at right angles to one another, simulating the seat and back of a chair. The upholstery fabric to be tested covers the exposed surfaces of the foam. A lit cigarette is placed in the crevice formed by the two foam pieces, and is then covered by a piece of a standard lightweight fabric. The test result is the mass loss of the assembly during the 45 min duration of the test. For the development of the SRM/PUF, NIST used a cotton upholstery fabric with consistent high smoldering.

- A box test method that measures the tendency of a padding material to smolder, with no complications from possible variability in the upholstery fabrics and the assembly of the composite. A slab of foam is placed in a tight-fitting plywood box and cartridge heater is inserted into the center of the foam. Holes bored into the plywood provide a path for lateral air flow, while a piece of glass fiber cloth placed on top of the foam slab prevents vertical, buoyancy-driven air flow at the cavity formed around the cartridge heater during the test. Thermocouples are placed in the foam at several distances from the heater to measure the temperature rise. The heater is raised to a temperature high enough to initiate smoldering of the foam. "Self-sustained smoldering" is achieved when all the thermocouples reached temperatures above 100 °C. The test is of 100 min duration. Also measured are the onset temperature for self-sustained smoldering, the mass loss of the foam, the diameter of the smolder zone, and the power supplied to the heater.

- The open-flame configuration method to be used in the CPSC proposed standard. The purpose of this test is to evaluate the effectiveness of barrier materials. Thus, the PUF needs to be prone to substantial mass loss from flaming ignition. There are two parts to this testing.

  - Bare foam test. This involves a two-piece mockup, similar in shape to the first test described above, but with larger foam slabs. A small gas burner is applied to the crevice for 5 s. The test is concluded at 120 s or when the mass loss reaches 20 %. If the mass loss is lower, the PUF is not usable for the second part of the test.
  - Composite test. This test uses the same type of foam mockup as the bare foam test but the foam is covered with a barrier fabric and a standard upholstery fabric. A larger flame impinges on the crevice for 70 s. The mass loss is recorded for 45 min or until no further mass loss is observed, whichever occurs first.

- The measurement of flame spread rate method. A single, horizontal slab of the foam is laid on a drywall plate, which is located on a load cell. The foam is ignited along one edge by a gas burner. The flame spread rate is determined using the measured time for the flame front to move a fixed distance. The mass loss rate is determined using the change in specimen mass that occurred during the flame spread interval. This test uses less foam than the above described tests.

The experimental design was organized into four parts or Iterations as follows.

- Iteration 1: Evaluation of the effect of the chemistry of the foam. Five polyether polyols in combination with three surfactants were examined. All five polyols were reacted with a single mixture of isomeric toluene diisocyanates. A common mixture of catalysts and emulsifiers was used, and the blowing agent was deionized water. There were considerable differences in foam density and air permeability among the foams.

  The mass loss in the box tests indicated no correlation with foam density, air permeability, catalyst content, the relative humidity during the test, or the air flow above the enclosure. The onset temperature for smolder was between 320 °C and 340 °C for all the formulations. The average mass loss for all the formulations at a temperature of 340 °C was the same as for a temperature of 360 °C. There was no effect of the surfactant type on the mass loss. For eight of the formulations, there was no effect of the polyol type on mass loss. For two chemically identical batches of the ninth formulation, there was a factor-of-two difference in mass loss; however, the two batches were foamed at different temperature and relative humidity. Examination of the lower smoldering batch showed some formation of closed cells creating lower air permeability. Examination of the remaining two foams indicated larger pore sizes and thus a reduction in internal surface area. Combined, these results indicated at most a modest dependence of smoldering mass loss on foam chemistry (within the range of formulations) and a potential dependence on air permeability and internal surface area.

The fire spread rate (FSR) measurements showed that the foam formulation had no significant effect on fire spread rate. The mass loss rate (MLR) values in the same tests showed no dependence on the polyol, but there was a systematic dependence on the surfactant. Surfactants with flame-retardant properties may suppress flaming over the liquid produced by thermal degradation of the foam. This might have an impact on the open-flame test to be used in the CPSC proposed standard.

- Iteration 2: Evaluation of the effect of processing parameters. For a single polyol/surfactant formulation, a full factorial experimental design with high and low levels of water, tin catalyst, and mixing head pressure produced flexible foam samples with a range of densities (28.4 kg·m^{-3} to 34.3 kg·m^{-3}) and air permeabilities (3.1 m·min^{-1} to 74.2 m·min^{-1}).

An air permeability value above 70 m·min^{-1} was a required specification, but not necessarily sufficient to achieve a mass loss of 25 % in the mockup test. In this high range of air permeability, the data suggested that at least one other variable was impacting smoldering. Thus, a better morphological description of the foam structure was required.

The highest air permeability and smoldering in these specimens was achieved at low tin catalyst and low pressure levels.

- An increase in head pressure or tin content correlated with a decrease in smoldering.
- An increase in water content correlated with an increase in smoldering in the box test and a decrease in the mockup test.
- An increase in head pressure or tin content correlated with a decrease in permeability, while an increase in water content correlated with an increase in permeability.

The cell size increased with increasing mixing head pressure and higher water content. At the higher water level, the cell size decreased with the tin catalyst content.

- Iteration 3: Evaluation of the effect of cell morphology. The reference foam used for this Iteration was the foam from Iteration 2 which had the highest air permeability. A second, identical formulation was produced at a lower mixing-head pressure because it was found that the head pressure is the main processing parameter controlling smoldering at high water level. A third foam was produced the same way as the first, except that it did not contain the processing aid.

The mass loss values in the mockup tests of the three foams were very different despite having similar air permeability values. This suggested that there may be other parameters influencing smoldering. It was proposed that two foams could have significantly different pores sizes and ratio of open/closed membrane and still yield the same air permeability.

Air access to the pores is critical. A high fraction of closed cells retards air movement through the foam. The surface-area measurement does not discriminate between closed and open cells, since krypton, the gas adsorbed in the measurement technique, can diffuse through the membranes. Thus, a specification for a smoldering foam must supplement a surface-area measurement with information from a micrograph or an air permeability measurement.

For foams of high air permeability, as the internal surface area of the foam increased, the smoldering mass loss in the mockup test increased. As the cell cross-sectional area decreased, the smoldering mass loss increased. Holding the foam mass density constant, having smaller pores means having more pores and thus more internal surface area at which the smoldering reaction occurs.

- Iteration 4: NIST is collecting, analyzing, and interpreting characterization and smoldering data of PUFs produced from the foamer's production/manufacturing line. The data (which will be provided in a separate document) indicates it is possible to commercially produce foam with consistent smoldering.

After working with an experienced foamer and performing extensive testing, NIST has come to the following conclusions.

At the pilot scale, it was possible to make high-smoldering foam targeted to a particular smoldering mass loss using the proposed test method. However, the intrinsic low repeatability and limited size of the bun (block of foam obtained in a single foaming process) in the pilot scale would generate relatively high costs. Because more buns would be needed to generate a given volume, and because bun to bun variability is significant, a pilot plant is less than ideal in terms of the variability/quality of a final potential SRM.

On a commercial production line, it may be feasible to produce a polyurethane foam with an approximate 30 % smoldering mass loss provided that the manufacturer can meet specifications for chemical formulation, air permeability, average cell cross-sectional area and mass density. However, it appears that on a manufacturing scale it may be difficult to achieve these specifications in a reproducible way within and between buns. Identifying conforming sections of a bun would require extensive testing, time and resources.

Using the current manufacturing technologies, it may be feasible to produce a high smoldering foam with lower variability within the bun, but with a smoldering mass loss value closer to 20 %. Since this mass loss value is less than the targeted 30 %, there is little information on the bun to bun variability and validation of the values for the targets parameters. The target for this type of foam should meet the following criteria:

- Pure polyurethane foam based on formulation C1 (see the main text for more detail), with no significant visible defects or impurities, in which the processing parameters (amount of catalysts, water and head pressure) are adjusted to meet the criteria reported below for air permeability, cell size and mass density (uncertainties equal to one standard deviation).
- Air permeability: (71 ± 8) m·min^{-1}
- Average cell area: (0.31 ± 0.01) mm^2
- Mass density: (27.4 ± 0.2) kg·m^{-3}

Additional effort will be required to identify a manufacturer with quality controls sufficient to produce a long-term supply of foam consistent with CPSC's proposed test standard.

List of Tables

List of Figures

List of Acronyms, Abbreviations, and Symbols

2-D	2 Dimensional
Φ	Air permeability
$\Phi_{threshold}$	Threshold value for air permeability
Σ	Mean values of area per cell calculated in a selected cross-section
Avg	Average value
atm	Standard atmosphere (pressure)
BET	Method used for the calculation of surface area
BS	British Standard
CFM	Cubic feet per minute
CFR	Code of Federal Regulations
CPSC	Consumer Product Safety Commission
Dev	Average of the absolute deviations of data points from the mean value measured
Dev%	Average relative deviation expressed as a percentage of the mean value measured
EDM	Euclidian distance map
FSR	Flame spread rate
GA	State of Georgia
IFD25	Indentation force deflection 25 %
IFD	Indentation force deflection
MA	Commonwealth of Massachusetts
MD	State of Maryland
ML_{120s}	Mass loss measured in % by mass 120 s after ignition during the open-flame mock up test
$ML_{340+360}$	Average mass loss percentage measured in the box test at a heater temperature of 340 °C and 360 °C
ML_{mockup}	Mass Loss percentage measured in the cigarette smoldering mockup test
MLR	Mass loss rate
NIST	National Institute of Standards and Technology
NY	State of New York
PA	Commonwealth of Pennsylvania
php	Parts (by mass) per hundred parts of polyol for a specific component in a foam formulation (*e.g.*, 10 php of X means that 10 g of component X are used in combination with 100 g of polyol).
psi	Pound-force per square inch
PUF	Flexible Polyurethane Foam
R.H.	Relative humidity
RM	Reference Material
SRM/PUF	Standard Reference Material for Flexible Polyurethane Foam
SFPE	Society of Fire Protection Engineers
SRM	Standard Reference Material
SSA	Specific surface area (area per unit volume)
StDev	Standard deviation
StDev%	Relative standard deviation expressed as a percentage of the mean
TC	Thermocouple

| T_{max} | Maximum temperature measured 102 mm away from the center of the cartridge heater during the box test |
| U.S. | United States |

This page is purposely left blank.

1. Introduction

A review of fire statistics from the last two decades shows a reassuring decline in home fire deaths.[1] Yet despite decline, upholstered furniture and bedding remain the most frequent "first items to ignite" that result in residential fire deaths in the United States.[2] According to estimates by the United States (U.S.) Consumer Product Safety Commission (CPSC), a large number of these fire deaths can be attributed to smoldering materials commonly found in upholstered furniture and bedding. Despite the promising introduction of Reduced Ignition Propensity cigarettes in all 50 states[3], smoldering of upholstered furniture and bedding remains a threat to life and property.

Initiation of smoldering furniture fires is a complicated process that depends on the material properties of the ignition source (*e.g.*, smoking material), the fabric, the filling, the interactions between the materials, and the design of the upholstered furniture. To date, the propagation mechanism of smoldering in furniture, which is composed of textile fabrics and porous fuels, such as polyurethane foam (PUF), remains largely unstudied. The variety of material formulations and the intrinsically low reproducibility of the foaming process have effectively prevented a thorough understanding of the properties controlling smoldering in polyurethane foams. Ihrig *et al.* came the closest to explaining the impact of polyurethane substrates on the probability of smolder initiation by determining that the air permeability of the polyurethane foams was the overriding material property controlling smoldering propensity.[4,5] Unfortunately, no further analysis or morphological description of the foams was given. Empirical evidence showed that air permeability was very important to smoldering, but there was not enough information provided to fully explain the differences in the smoldering propensity of different foam formulations.

The factors affecting open-flame flammability in flexible polyurethane foam have been extensively studied; however, quantification of the key properties that determine smoldering propensity of such materials has been largely neglected.[6,7] This is mainly due to the inability to procure foams with consistent and homogeneous properties. Ohlemiller and Rogers described the effect of the fundamental properties of porous fuels on smoldering and attempted to better understand the relationship of smoldering to PUF properties; however, their work was limited because of difficulties in obtaining foam samples produced under well-controlled conditions.[8,9,10]

A great body of evidence has been established through numerical simulation of smoldering combustion of PUF and indicated the significance of air permeability on the rate of smolder propagation.[11,12,13,14] Thermal analysis of the foams has been performed in great detail in order to obtain multi-step models of foam pyrolysis and char oxidation that provided input data for models.[13] However, the relationship of the input parameters and actual foam formulation parameters has never been established. In many studies, morphological description of the PUF has been limited to the simplest terms.

The purpose of the research presented here was to understand the effect of basic formulation parameters and foam morphology on the initiation and extent of smoldering in PUF. The research then evaluated whether PUF with reproducible and well characterized smoldering could be used to produce a reference material (RM) for ultimate use in a standard test. For this purpose,

PUF was produced under controlled conditions and its smoldering propensity was evaluated using two test methods:
- The upholstery cover fabric smoldering ignition resistance test. This is a well-established test described in proposed CPSC 16 CFR 1634[15] (herein referred to as "CPSC proposed standard");
- The semi-quantitative test method. This is a new, robust test developed at NIST (the box smolder test, described in Section 2.4.1 of this report).[16]

The experimental design was organized in four parts, (referred to here as Iterations):
- Iteration 1: evaluation of the effect of **raw materials**;
- Iteration 2: evaluation of the effect of **processing parameters**;
- Iteration 3: evaluation of the effect of **cell morphology** in open-cell PUF;
- Iteration 4: **scale-up** from pilot plant to a production line.

Each Iteration started with material/formulation selection and sample preparation (*e.g.*, foaming, cutting and conditioning). The samples were then conditioned and tested. Finally, the output data were analyzed for selecting the PUF specifications for the next Iteration. At the end of this process, a potential SRM for PUF (SRM/PUF) was specified (Figure 1).

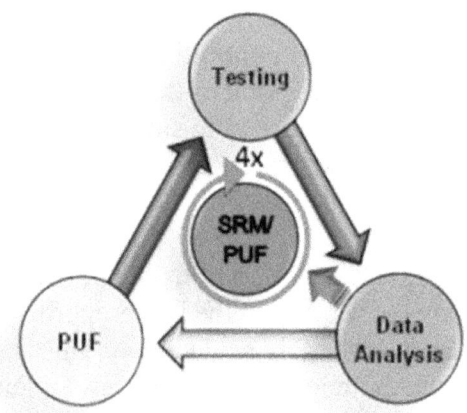

Figure 1. Schematic of the four-iteration experimental design for the SRM/PUF specification process.

The aim of the Iterations was to separately evaluate, in a controlled fashion, the effect of a single set of parameters (raw materials, processing parameters, or cell morphology) on smoldering. These parameters are not independent of each other; changing a single parameter may precipitate a change in another parameter. For example, a variation in raw materials may require an adjustment in processing parameters and/or induce a different cell morphology. Furthermore, climatic conditions (temperature, humidity and atmospheric pressure) during manufacturing may impact the cell morphology (cell size, strut thickness and length, open *vs.* closed cell structure, etc.) so that fine-tuning of processing parameters (water content, catalyst content and mixing head pressure[I]) is required on a daily basis to ensure consistent foam quality. This implies that,

---

[I] Mixing-head pressure is the pressure inside the mixing head. See Section "Sample preparation" for more details.

even for a given formulation, a bun-to-bun variability in the foam properties is somewhat inevitable for typical manufacturing systems. Significant variations in PUF properties were observed in the same foam bun (in-bun variability) as a function of the specific location due to surface-proximity effects (a relatively high density layer, the so-called "skin", is generated on the surface of a bun during foaming), temperature gradient effects (core-to-periphery decrease in the temperature of the bun) and gravity (top-to-bottom pressure increase).

The aforementioned phenomena, coupled with the intrinsic complexity of smoldering mechanisms, outline the challenges that need to be addressed for developing a potential SRM/PUF with a reproducible and well-characterized smoldering behavior.

2. Experimental Description[II]

This study required NIST to procure foams with specified chemical and morphological properties. Foam samples were custom made by a commercial foamer, according to NIST specifications. The materials used for preparing these PUF samples are listed below.

2.1. Materials

All materials were used as-received unless otherwise indicated. Five different polyether polyols (P1-P5) were selected for investigating the effect of polyols on smoldering. Their specifications, as provided by the manufacturer, are listed in Table 1. Similarly, three different surfactants (S1-S3) were selected for investigating the effect of surfactants on smoldering. Their specifications, as provided by the manufacturer, are listed in
Table 2. The other reagents used in the custom formulations were an isocyanate, water, catalysts and processing aids. Their specifications, as reported by the manufacturer, are shown in Table 3. Stannous octoate is referred to hereafter as "tin catalyst" or "tin".

2.2. Sample preparation

Samples were prepared in a small pilot plant or in a production line. In both cases, all reagents were pumped at a controlled rate into a fixed mixing chamber (mixing head). The pressure in the mixing head was adjusted by controlling a valve at the outlet. In the pilot plant, the material was transferred from the mixing head to a foaming box through a feeding tube. After 15 minutes at room temperature, the foams were cured in an oven at 110 °C for one hour and post-cured at room temperature for an additional 24 hours. In the production line, the ingredients of the foam formulation, discharged through the nozzle of the mixing head, fell onto the front of a conveyor belt. The temperature of the bun reached typically about (150 to 170) °C in water-blown foams. Curing was completed in air, and no post-curing was required. Samples were cut with an automatic laser system. All samples were conditioned at a temperature of (21±3) °C and between 50 % and 66 % relative humidity for at least 24 hours prior to testing.

[II] The policy of NIST is to use SI units of measurement in all its publications, and to provide statements of uncertainty for all original measurements. In this document however, data from organizations outside NIST are shown, which may include measurements in non-SI units or measurements without uncertainty statements.

Table 1. Polyol specifications provided by the manufacturer (uncertainties not provided).

Specifications	P1	P2	P3	P4	P5
	Polyether triol with ethylene oxide end cap	Polyether triol without end cap	Polyether triol produced by double metal cyanide catalyst	Bio-based polyol	Polyether polyol with copolymerized styrene and acrylonitrile
Molar mass[1] (g·mol^{-1})	3,000	3,200	3,000	1,700	-
OH number[1] (mg KOH·g^{-1})	54.5-57.5	50.5-53.5	54.5-57.5	56	31.1
Acid number[1] (mg KOH·g^{-1})	<0.02	<0.02	<0.02	<0.50	-
Water content (%)[1]	<0.05	<0.05	<0.05	<0.05	<0.07
Viscosity[1] at 25 °C (mPa·s)	480	520	580	3,200	4,750
Density[1] at 25 °C (kg·L^{-1})	1.02	1.02	1.02	1.00	1.04

Table 2. Surfactant specifications provided by the manufacturer (uncertainties not provided).

Specifications	S1	S2	S3
	Alkyl-pendant organo/silicone surfactant designed for low flammability PUF, blown with water and/or methylene chloride in a conventional flexible slabstock process	Organo/silicone surfactant designed for supercritical-CO_2-blown slabstock PUF and reduced flammability	Conventional polysiloxane block copolymer surfactant for slabstock and molded PUF
Viscosity at 25 °C (mPa·s)[1]	750	945	1,150
Density at 25 °C (kg·L^{-1})[1]	1.04	1.03	1.05

Table 3. Other components used in PUF by the manufacturer (uncertainties not provided).

Components	Description
TDI	Mass ratio mixture of 2,4- (80%) and 2,6-isomers (20%)
Deionized water	Blowing agent
Tin catalyst	Stannous salt of ethyl-hexanoic acid (stannous octoate)
Amine catalyst	Mass ratio of triethylene diamine (33%) and dipropylene glycol (67%), soluble in water and polyol.
Polyether catalyst	Polyether based catalyst
Processing aid	Mixture of emulsifiers based on a fatty ester used as a processing aid to decrease the viscosity; soluble in the polyol and insoluble in water.

2.3. Sample identification

The properties of a foam sample may vary depending on the specific formulation, the foam bun and the location in the bun; thus, it is useful to specify these parameters with an identification code for each custom made sample. The formulations are identified by a code as X# with X=A,B,C,D and the symbol '#' a number between 1 and 11. Formulations 'A' are formulations prepared for Iteration 1, formulations 'B' for Iteration 2, 'C' for Iteration 3 and "D" for Iteration 4. The symbol '#' indicates the formulation number in a given Iteration; '1' indicates the first formulation prepared for a given Iteration, '2' the second one, etc. For example, C2 and A11 indicate the 2nd formulation from the 3rd Iteration and 11th formulation from the 1st Iteration, respectively. A generic formulation X# can be foamed multiple times. The specific foam bun for a given formulation is identified by a number after the formulation code. For example, C2-14 indicates the 14th foam pouring from formulation C2. The foam location in a specific bun is indicated by a letter after the foam bun code; 'S', 'M', and 'E' are used to indicate samples that are cut from a region close to the pouring start position, middle position or end position, respectively. For example, C2-14S indicates a foam sample from formulation C2, 14th pouring, and cut from the pouring start region. For the samples prepared in the production line, a lower case letter (t, m, b) is also used to indicate samples that are cut from a region close to the top (t), middle (m) or bottom (b) of the bun. For example, D2-1Sb indicates a foam sample from formulation D2, 1st pouring, and cut from the pouring start region at the bottom of the bun.

A detailed description of foam properties, pouring conditions, and date of foaming for each bun is provided in Appendix 1.

2.4. Flammability Testing

The smoldering of PUF was evaluated by two tests. The first one, the Mockup Test (described in the CPSC proposed standard), aims to mimic a realistic ignition scenario for upholstered furniture; the second one, the Box Smolder Test (see Section 2.4.1) is designed to measure smoldering in a fire scenario with a setup that is not affected by variability in the testing materials (*e.g.*, barrier fabric, cotton sheeting, smoking material) other than the foam itself. Both of these tests provide information regarding the smolder propensity of the foam under different configurations.

In this project, formulation and processing of PUF were experimented targeting a specific smoldering behavior, suitable for the proposed CPSC standard. In doing so, the possible implications for open-flame configuration, which are also part of the CPSC proposed standard, needed to be considered. For this reason, even though smoldering propensity was the main focus of this study, open-flame combustion was also investigated. Open-flame resistance was measured by two methods: the *Mock-up Open-flame Test*, described in the CPSC proposed standard, and *Flame Spread Test*, a newly developed method intended as a smaller scale alternative to the *Mock-up Open-flame Test*. Both methods are discussed in detail below.

2.4.1. Box Smolder Test

The 'box' in the Box Smolder Test is a newly developed apparatus for assessing smoldering in PUF.[16] A heater (a copper cylinder, 19 mm in diameter and 95 mm in length with a concentrically inserted cartridge heater, 6 mm in diameter and a maximum power of 85 W) is inserted into the core of a ($280 \times 280 \times 153$) mm^3 foam sample contained in a plywood box with 13 mm thick walls (Figure 2).

Three holes, 13 mm in diameter were cut into each side of the box to allow lateral air flow into the foam block. The top surface of the foam was covered by a glass fiber cloth (planar density of 558 g·m^{-2} ± 12 g·m^{-2}) in order to prevent chimney effects due to cavities formed in the foam block around the cylindrical heater. The glass fiber cloth also served to minimize the risk of transition to open flaming by reducing the oxygen supply.

The temperature of the heater was increased from room temperature to a set temperature between 320 °C and 360 °C at a heating rate of 6 °C·min^{-1}. The temperature in the foam was monitored with six thermocouples (0.5 mm thick K-type thermocouples, KMQSS-020G-6, Omega Engineering Inc.) at distances of 50 mm (TC1 and TC4), 75 mm (TC2 and TC5) and 102 mm (TC3 and TC6) from the center of the heater. The thermocouples were held by 47 mm long syringe needles with the exposed thermocouple tip positioned at a height of 57 mm from the bottom of the plywood box. A thermocouple was also used to monitor the temperature of the heater; the power necessary to keep the temperature of the heater at the target temperature was recorded (Figure 3). The cylindrical heater was powered by a custom fabricated power supply with a proportional–integral–derivative controller (model CN77343, Omega Engineering Inc.). The power supplied to the heater was recorded using a power meter with an accuracy of ± 0.1 W·h (*Watt's Up?* pro 99333, Electronic Educational Devices). Thermocouple readings were recorded with either of two types of digital data loggers (OM-Daqpro-5300, Omega Engineering Inc. and CR23X, Campbell Scientific Inc.).

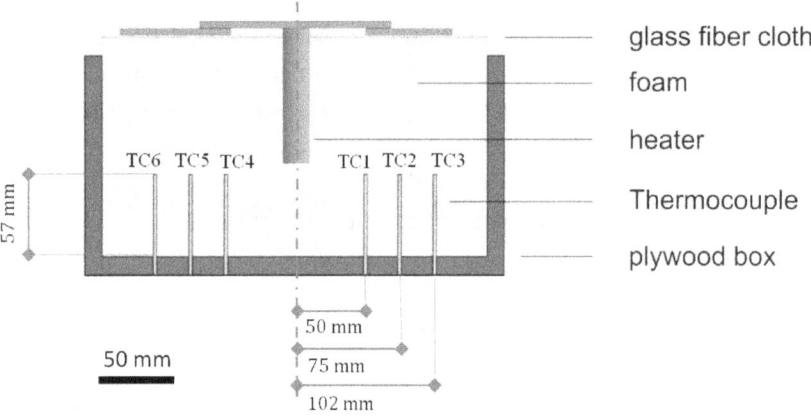

Figure 2. Schematic drawing of the box test setup, showing the test specimen and the location of the thermocouples.

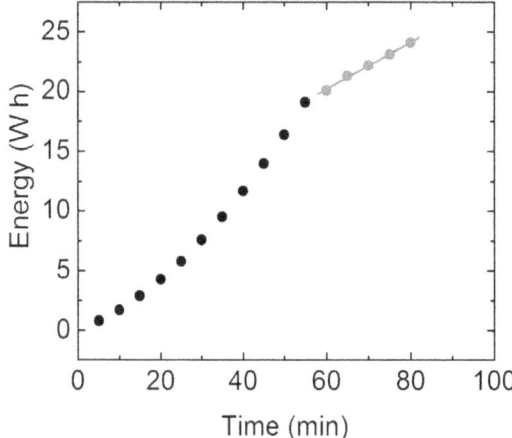

Figure 3. Energy consumed by the heater to reach and sustain a set temperature of 360 ºC. The black points indicate the energy consumption during the heating; the red points indicate the energy input at the set temperature. The red line is a least-squares regression fit to the red data points and its slope is used to calculate the average power necessary to stabilize the heater at the set temperature.

Two smolder box setups were run concurrently. Both boxes were contained in a custom fabricated enclosure with inner dimensions of $(165 \times 86 \times 66)$ cm^3 and open top (Figure 4). A perforated aluminum grid was installed 280 mm above the counter to ensure sufficient ventilation by natural convection, and the smolder boxes were placed on a platform, 85 mm above the grid. An air flow of (19 ± 4) m$^3 \cdot$min^{-1} measured at the top rim of the enclosure was maintained by a canopy hood positioned atop the enclosure. Lower levels of ventilation could not be used since a removal of the toxic exhaust gases was required. The heater was set to maintain a specified temperature until it was switched off 80 min after the start of the test (t=80 min). The thermocouple measurements were continued for an additional 20 min, for a total of 100 min. The test was then terminated and the mass loss of the foam was determined by weighing the sample with the box. If smoldering persisted, the sample was quenched after the mass determination by adding water with a spray bottle equipped with silicone tubing and a syringe needle.

The box test was designed to investigate foam smoldering in a well-characterized fashion by measuring the following set of parameters:
- mass loss,
- temperature profiles,
- diameter of smolder zone,
- power supplied to the heater, and
- onset temperature for self-sustained smoldering.

Mass loss was measured at the end of the test (t=100 min) by measuring the mass of the foam sample in the box without removing the char that was produced by smoldering. This is a major difference between the box test and the CPSC proposed standard (upholstery cover fabric smoldering ignition resistance test), where all char is removed and included in the mass loss.

Figure 4. Enclosure for performing box tests.

An example of temperature profiles for two commercial foams is shown in Figure 5. As discussed above, the heater is turned off at t=80 min, once the heating ramp and isothermal stage are completed, and data are collected for an additional 20 minutes after turning off the heater, resulting in a 100 minute total test duration. In the measurements shown in Figure 5, the heater was tuned to a set point of 360 °C. The temperature of the smolder-prone foam continued to rise in the outer parts of the sample independent of the heater temperature. Comparing this result to a sample that did not smolder intensively demonstrated that the temperature rise was due to the smoldering reaction and not thermal diffusion from the heater into the sample. Temperature rise throughout the foam block is an indicator of smoldering intensity.

A self-sustained smoldering foam is defined in the context of this test as a foam that reaches a temperature above 100 °C for all six thermocouples at the end of the box test. In self-sustained smoldering, generally, the heater temperature drops well below the temperature of the smolder zone. The foam continues to smolder on its own, as indicated by smoke generation, and the sample has to be extinguished with water. Samples that did not exhibit sustained smoldering had a mass loss below 2 %. Figure 6 shows the correlation between mass loss and average temperature for the six thermocouples reached at the end of the box test.

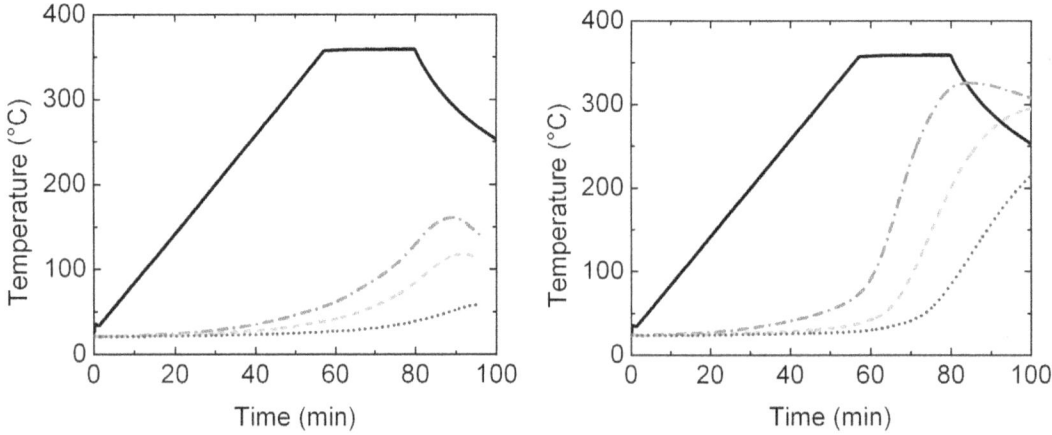

Figure 5. Temperature profiles measured in the smolder box test for a foam showing self-sustained-smoldering (right) and a foam in which smoldering is not self-sustained (left). The temperature of the cylindrical heater is shown as a solid line (target temperature of 360 °C). The averages of two thermocouple temperatures measured in the foam are shown for distances from the heater of 50 mm (dash-dot), 76 mm (dash) and 102 mm (dot).

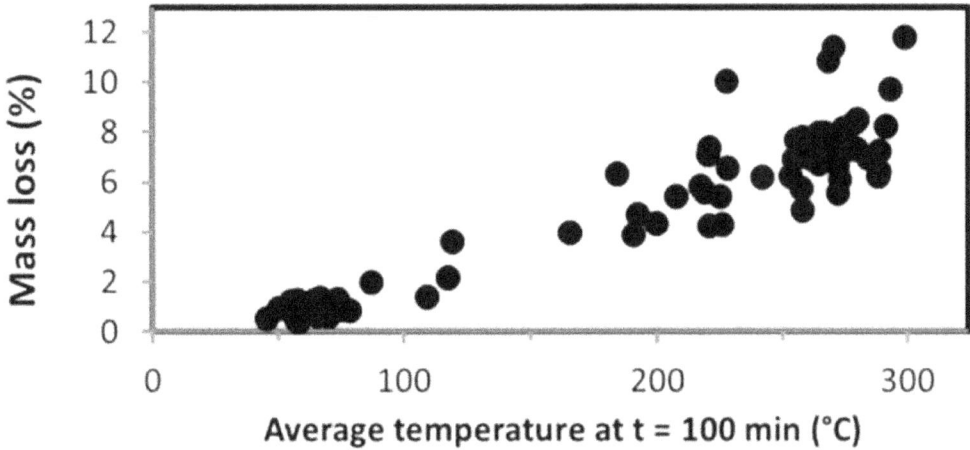

Figure 6. Comparison of mass loss and average temperature of the six thermocouples at the end of the box test. Data are collected on foams from Iteration 1 and 2.

The diameter of the smolder zone was measured by cross-sectioning the foam residue. After the sample had cooled down and dried, if extinguished with water, the foam block was cut into two parts and the diameter of charred material inside the foam was determined (at about mid-depth). A visual comparison of the residue of samples that smoldered with different intensity is shown in Figure 7. Even though the mass loss was only a single digit percentage of the sample mass, the diameter of the charred region was substantial for smolder-prone samples. The diameter of the smolder zone was only measured on few representative samples from Iteration 1.

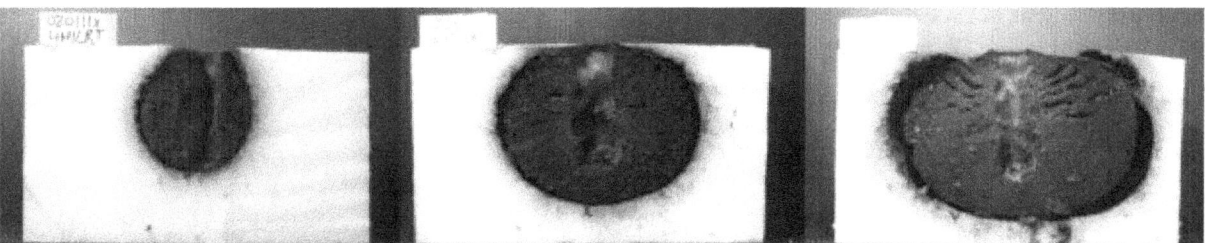

Figure 7. Illustration of the charred material produced in the box test. The respective mass losses were 2 % by mass (left), 4 % by mass (center) and 7 % by mass (right).

Power supplied to the heater can also be used to assess the intensity of the smoldering reaction. Figure 8 shows the mass loss in the box test versus the power required to maintain the temperature of the heater at the target temperature during the test. For all samples that attained a high mass loss, the power supplied to the heater was much lower than for samples that did not significantly lose mass. This established that the heater was not driving the temperature rise in smoldering samples.

The smolder box test was also designed to determine the onset temperature for self-sustained smoldering in the PUF samples. The onset temperature for self-sustained smoldering is defined as the lowest set temperature of the heater at which the foam shows self-sustained smoldering. The onset temperature was investigated by conducting box tests at four different target temperatures of the heater (320 °C, 330°C, 340 °C, and 360 °C), and observing the lowest temperature at which sustained smoldering was achieved.

The box test provides a characterization of smoldering propensity in a specific fire scenario. There is no direct relationship to all possible fire scenarios. The presence of the box itself reduces smoldering, likely by limiting the buoyant convection (*i.e.*, oxygen supply).[III] Therefore, the mass loss data reported here apply to this specific setup and cannot be considered intrinsic material properties. Modifications to this testing device and protocol are currently being considered (e.g.; improving air flow by removing the wooden sides and bottom). Initial experiments without the box sides have shown this modification significantly increases mass loss.

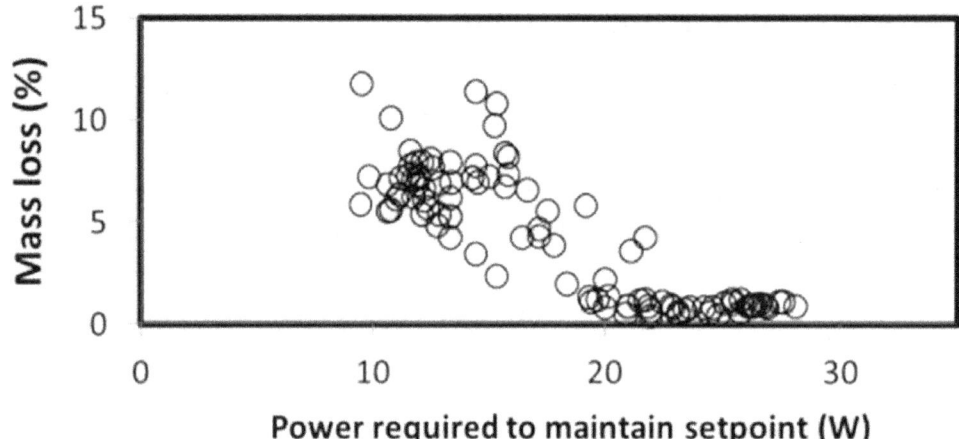

Figure 8. Plot of total mass loss in the smolder box test versus the power required to maintain the temperature of the heater at the target temperature. For samples that intensely smoldered resulting in high mass loss, only low power was needed for the heater. Data are collected on foams from Iteration 1 and 2.

2.4.2. Mock-up Smolder Test

This test method was designed to measure the resistance of an upholstery cover fabric to a smoldering ignition source when the fabric is placed over a standard polyurethane foam substrate. The objective here was to develop a PUF with well characterized and reproducible smoldering to be used as a standard substrate for fabric testing. To accomplish this, the mockup test was used to measure smoldering of a specific PUF sample for a specific upholstery cover fabric (100 % cotton, indigo twill weave and an average mass of 0.68 kg·m^{-2})[IV].

For a full description of the operation procedures and test set-up refer to the CPSC proposed standard.[15] Contained here is a brief description with a schematic drawing of the test setup (Figure 9) and a set of foam mockups that are ready to be tested (Figure 10). The mass of the

---

[III] A comparison of the mass loss of a foam formulation measured with a box (three replicates) and without a box (three replicates), showed that the mass loss increased by a factor of two when a box was not used.

[IV] Average mass data measured by the manufacturer.

$(203 \times 203 \times 76)$ mm^3 vertical and $(127 \times 203 \times 76)$ mm^3 horizontal PUF is recorded and is later used in calculating the smoldering mass loss value. The PUF is covered with the upholstery fabric (Indigo Buckaroo Denim purchased from Jo Ann Fabrics was used in this study), and the entire ensemble is placed in the specimen holder. A lighted cigarette (Standard Cigarette for Ignition Resistance Testing, NIST SRM 1196)[17] is placed in the crevice formed by the intersection of vertical and horizontal panels of each test assembly. Each cigarette is covered with a piece of sheeting fabric, a 100 % cotton, white plain weave of (19 to 33) threads/cm^2, and a weight of (125 ± 28) g·m^{-2}. The cigarette is allowed to burn its entire length. After 45 min, the PUF (char and non-char) is separated from the other testing components (e.g., cigarette ash and fabric) and the mass of this PUF is measured. The char is removed and the mass of the remaining non-charred foam is measured. The smoldering mass loss value is calculated as the difference of between the non-charred PUF mass and the original PUF mass divided by the original PUF mass.

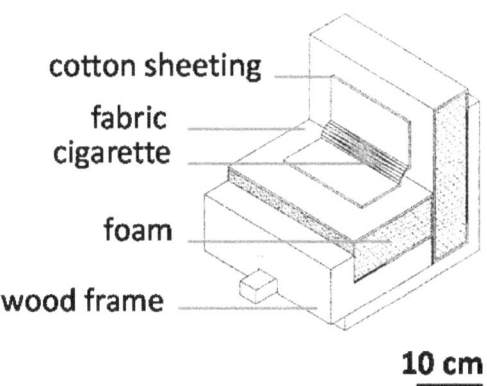

Figure 9. Schematic drawing of the mockup test described in the CPSC proposed standard (upholstery cover fabric smoldering ignition resistance test).

Figure 10. A set of foam mockups ready to be tested prior to the addition of the lighted cigarette.

2.4.3. Mock-up Open-flame Test

The open-flame resistance was measured according to the CPSC proposed standard.[15] This proposed standard includes open-flame tests on foams with or without a barrier fabric (bare foam) that uses a mock-up assembly similar to BS 5852.[18] Two blocks of foam per mockup

11

assembly are used. The vertical block is (457±5) mm x (305±5) mm x (76±2) mm; the horizontal block is (457±5) mm x (83±5) mm x (76±2) mm. An example of a mockup-assembly without a barrier is shown in Figure 11.

The open-flame test setup for bare foams (without barrier) used here is described in the CPSC proposed standard. It is intended to define the flammability performance requirements for standard polyurethane foam in an assembly of mockups. This test is a slab of bare foam (no cover or barrier fabric) impinged by a 35 mm butane flame for 5 seconds. According to CPSC proposed standard, 120 s after removing the open-flame the mass loss of the bare foam should be greater than 20 %. The bare PUF mockup assembly and frame was placed on a metal tray and mounted on a scale; the mass loss of the sample was measured and monitored in real time. The open-flame ignition source was applied to the seat/back crevice of the mockup, as seen in Figure 12. A hand-held carbon dioxide extinguisher extinguished the flames after 120 s or when the mass loss reaches 20 % by mass.

Figure 11. An example of a mockup assembly without a barrier used in the open-flame test for bare foam.

Figure 12. Photo of test with formulation C1 a few seconds before extinguishing the sample. At this stage, flaming liquid material is dripping to the underlying tray, but there is no pool fire (which would sustain flaming and boost the heat release rate through a feedback effect).

The open-flame test with barrier material is intended to measure the open-flame ignition resistance of interior fire-barrier materials to be used over standard polyurethane foam in a mockup assembly. This test is detailed in the CPSC proposed standard. Briefly, the interior fire-barrier material (a fabric 153 g·m⁻² product composed of a fiberglass base needle-punched with polyester and modacrylic fibers, covered with a polyester batting, nominally 153 g·m⁻², 9.5 mm thick, and nonwoven) is placed between a standard cover fabric (100 % bright regular rayon, scoured, 20/2 ring spun basket weave construction, (271 ± 17) g·m⁻²) and the foam, and assembled on a metal test frame.

The open-flame ignition source was a 240 mm butane flame, impinged onto the seat/back crevice of the sample for 70 s. The mass loss was recorded for 45 min or until the mockup self-extinguished and no mass change was observed, whichever occurred first. The ignition source at the beginning and end of flame impingement is shown in Figure 13 and Figure 14, respectively.

After ignition the flame propagated on the fabric covering the top vertical block of foam and the top side of the horizontal block. Flame-out occurred between 7 min 30 s and 15 min, but the samples continued smoldering for several minutes more, inducing further mass loss.

Figure 13. A photo of the 240 mm butane flame impinging on the standard rayon material covering the foam for 70 s in the open-flame test with barrier material.

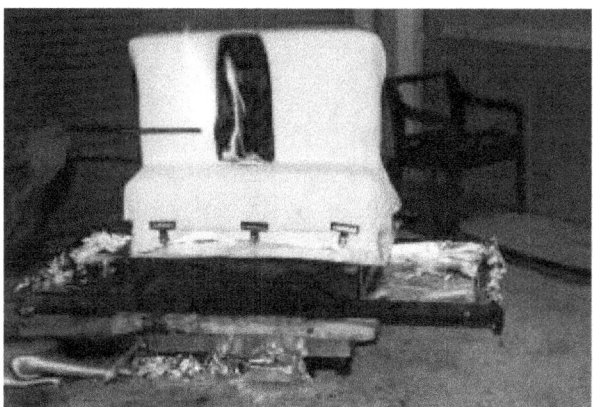

Figure 14. Photo of the open-flame test with barrier material. The 240 mm butane flame was removed from the crevice after 70 s of flame impingement.

2.4.4. Flame Spread Test

Open-flame resistance was measured by the *Flame Spread Test*, a newly developed method intended as a smaller scale alternative to the *Mock-up Open-flame Test*. The setup is shown in Figure 15. The foam sample $(28.0 \times 7.5 \times 3.0)$ cm^3, insulated underneath by a drywall plate, was placed on an aluminum-foil catch-pan to retain flaming liquid and prevent dripping. The sample mass was monitored in real time by a load cell. Ignition was achieved by impinging the samples for 20 s with a T-burner. Any pool developed during the tests was fully contained within the pan. The pool was immediately behind the flame front and was rapidly pyrolyzed. A video acquisition system was used to record the entire test and monitor the position of the flame front. The speed at which the flame front spread over the surface of the PUF (flame spread rate) was measured.

The average flame spread rate (FSR) was calculated by measuring the time (t_{spread}) required by the flame front to travel from a mark placed at 76.2 mm from the pilot ignited end to a location at 228.6 mm. An average mass loss rate (MLR) was also calculated for each test by dividing the mass loss measured at t= t_{spread} (designated as m_{spread}) by t_{spread}.

Sample size: 28 cm × 7.5 cm × 3 cm

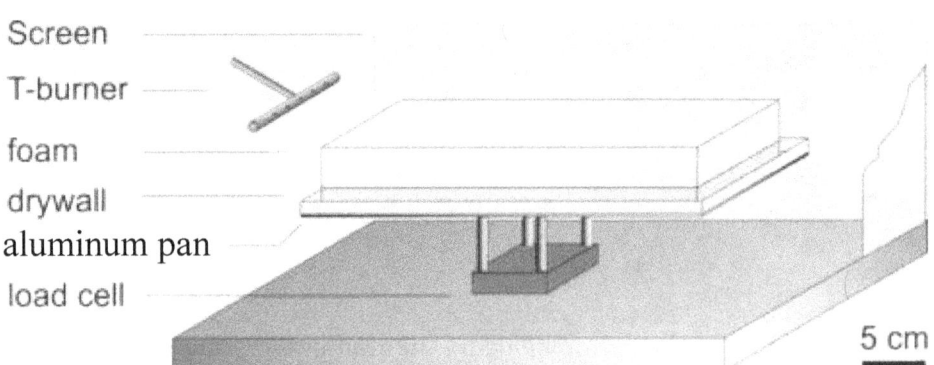

Figure 15. Schematic drawing showing the setup used for measuring flame spread.

2.5 Foam Characterization

Foams are three-dimensional structures containing gas bubbles (cells). In PUF, cells are polyhedrons, most commonly dodecahedrons with pentagonal faces, separated from each other by thin sections of polymer: the foam struts (polymer at the shared edge of the polyhedrons) and the foam membrane or window (polymeric thin film connecting the struts on a face of the polyhedron). An open-cell is a cell with only open windows, *i.e.*, no residual membrane. A closed-cell is a cell completely separated from the adjacent ones by windows. PUFs typically contain closed and open cells, as well as partially open cells (cell with few residual membranes).

The openness or porosity of PUF is generally described by measuring the air permeability. Briefly, the differential pressure is measured for a PUF sample with a given thickness and area as air flows through the sample. An electronic high differential pressure air permeability-measuring

instrument (FAP 5352 F2, Frazier Precision Instrument Co. Inc., Hagerstown, MD) was used in this study (
Figure 16). Foam was cut into samples (90x90x13) mm^3 and placed in a circular clamp, exposing 38.5 cm^2 to perpendicular air flow.

The target pressure-drop through the 13 mm thick foam slice was set to 127 Pa (13 mm of water). Nozzles with orifice diameters of 1.0 mm, 1.4 mm, 2.0 mm, 3.0 mm, 4.0 mm, 6.0 mm, 8.0 mm, or 11.0 mm were used in order to reach the target pressure drop. The test was conducted at room temperature. The value of permeability (Φ) in terms of volumetric air flow was read in cubic feet of air per square foot of sample area per minute (CFM) at about 20 °C and 1 atm and was converted to cubic meters per square meter of sample per minute (or simply meters per minute) at a temperature of 0 °C and a pressure of 100 kPa.

The air permeability for all samples prepared in the pilot plant was provided by the manufacturer. The value of Φ measured for each foam bun is reported in Appendix 1. Air permeability was measured by the manufacturer on a foam slice collected near the center point between pour start and pour end and at a depth of about 2.5 cm from the top surface of the bun. The foam slice was cut parallel to the bottom surface of the bun.

Air permeability measurements were also conducted by NIST to assess potential variations in permeability along the direction of pouring. Foam slices were cut perpendicular to the direction of pouring. Eight measurements per slice were performed.

The slices used for the NIST and vendor measurements were from planes orthogonal to each other. Due to anisotropy in the foam, cells looked roughly like circles in the slice used by the vendor and ovals in the slices used by NIST. This anisotropy also generated differences in the air permeability measured respectively by NIST and by the vendor.

The manner in which the sample is clamped is critical due to possible leaks. An example of a clamped PUF specimen is shown in Figure 17. This type of clamping was originally developed for textiles, but has proved to be more reliable than the concentric-cylinders mount (conventionally used for foams), likely due to improved sealing.

PUF foam firmness and load bearing capacity is described using a standard measurement value called IFD (Indentation Force Deflection). An IFD number represents the force in pounds that is applied by a circular indenter to induce a given deflection expressed as a percentage of the initial thickness of the PUF sample. The values of IFD reported in this study were measured by a Zwick IFD Testing Unit equipped with a circular indenter (area 323 cm^2) at 25 % deflection (IFD25) on PUF samples with a 10.2 cm thickness and a square base of (30.5 × 30.5) cm^2. The IFD25 value is reported for each formulation in Appendix 1.

The surface area of PUF was calculated by Brunauer-Emmett-Teller (BET) measurements[19] carried out by an experienced testing laboratory. The BET values of surface per unit mass of sample are then converted in mass per unit volume (specific surface area) dividing the BET values by the density of the sample. The specific surface area (SSA) is defined here as surface per unit volume and it is expressed in inverse meters. The samples used for BET measurements were about $(1 \times 1 \times 23)$ cm^3.

The thermal conductivity and thermal diffusivity were measured on three inch thick samples of foams by the transient plane source measurement technique, previously described in detail.[20]

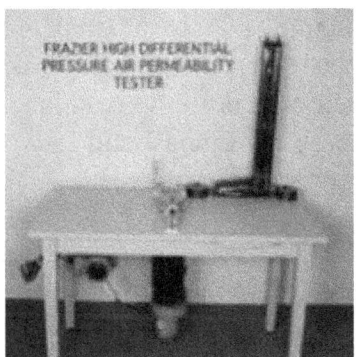

Figure 16. Photograph of the electronic high differential pressure instrument for measuring air permeability.

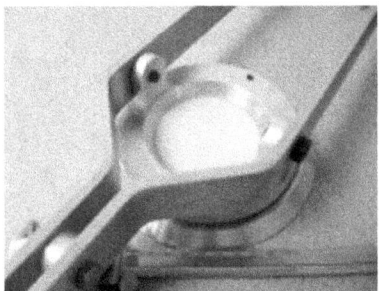

Figure 17. Clamping of the PUF samples during a permeability measurement.

3. Results and Discussion

3.1. Iteration 1: Impact of Polyols and Surfactants

The purpose of Iteration 1 was to determine if the chemical composition (within the parameters chosen for this project) impacts the PUF smoldering and open-flame behavior. The typical precursor components used in a PUF formulation are polyether polyols, a monomeric or oligomeric isocyanate, water, surfactants, catalysts, and additives (*e.g.,* fire retardants and colorant). Typical commercial grade PUF formulations were used and additives were excluded to avoid unnecessarily increasing the number of composition variables.

In Iteration 1 the smoldering performance of the foams was mainly evaluated with the box test because the repeatability in the mockup test may be affected by the variability in the mockup components (fabric, cotton sheeting, etc.).; The mockup test was used only on selected formulations to verify, as suggested by the box test results, that smoldering performance was dominated by morphology and not chemistry (within the experimental parameter space investigated in this project). The open-flame behavior of the foams was investigated by the small scale flame spread test described previously in order to limit the amount of foam prepared in Iteration 1.

3.1.1. Materials for Iteration 1

Polyols are the largest mass-fraction component in PUF. Polyether polyols are used in about 75 % of the global market for slabstock PUF.[21] This study investigated the effect of chemical structure of polyols on smoldering by using three polyether polyols (P1, P2, and P3). The three polyether polyols were all glycerol initiated, poly(propylene oxide) / poly(ethylene oxide) mixed block polyether triols. All three products had comparable OH numbers (52 $g \cdot mol^{-1}$ to 58 $g \cdot mol^{-1}$), similar molar masses (3000 $g \cdot mol^{-1}$ to 3200 $g \cdot mol^{-1}$), and acid numbers lower than 0.02 $g \cdot mol^{-1}$, according to the manufacturer. Polyols P1 and P2 differed by the existence of an ethylene oxide end cap for polyol P1. Both polyols were produced with a conventional process employing KOH as the catalyst. The third polyol P3 was produced in a double metal cyanide catalyzed process and had no ethylene oxide cap. Additionally, a bio-derived polyol (P4) and a graft-polyol (P5) were used as a possible replacement to conventional petroleum-based polyols. The specifications of polyols and all other materials used in this study are described in the Experimental Section.

The second largest component by mass fraction in PUF is the isocyanate. In Europe, PUF is often fabricated using a supercritical CO_2 process with diphenylmethane diisocyanate (MDI) and MDI oligomers. Outside Europe, a water based foaming process with toluene diisocyanate (TDI) is more common.[22] In this study, TDI was used exclusively to reduce the experimental parameter space.

Surfactants largely impact morphological-cell-structure parameters (*e.g.,* density, airflow, and surface area), which can play a significant role in the smoldering process. This study includes three silicone surfactants (S1, S2 and S3); S1 is a conventional silicon-based surfactant, and S2 and S3 are believed to impart some reduction in the severity of PUF flammability. Surfactant S2 is also commonly used in the supercritical CO_2 process.

Based on the above considerations, the impact of the chemical composition on PUF smoldering was assessed by characterizing 11 formulations containing five different polyols and three different surfactants. The experimental design is shown in Figure 18 with each dot representing a specific formulation (PiSj) for a given combination of polyol (Pi) and surfactant (Sj) type (i=1,2,3 and j=1,2,3,4,5 refer to polyol and surfactant type, respectively). In formulations containing the uncommon commercial polyols (P4 and P5), 25 % by mass of a standard polyol (P1) was replaced by P4 or P5. Formulations with all possible combinations of polyols P1, P2 and P3 and surfactants S1, S2 and S3 were foamed in order to investigate whether or not there was a systematic effect of either one of the components. A detailed description of all formulations of Iteration 1 (A1, A1R, A2, A3, A4, A5, A6, A7 A8, A9, A10, A11) is reported in Appendix 1, including type and quantity of reagents, atmospheric conditions during manufacturing, mixing head pressure, rise profile during foaming, air permeability, density, and indentation factor (IFD25). Formulation A1R was a repetition of A1 at a different temperature and humidity. Four replicate buns were foamed for each formulations of Iteration 1.

17

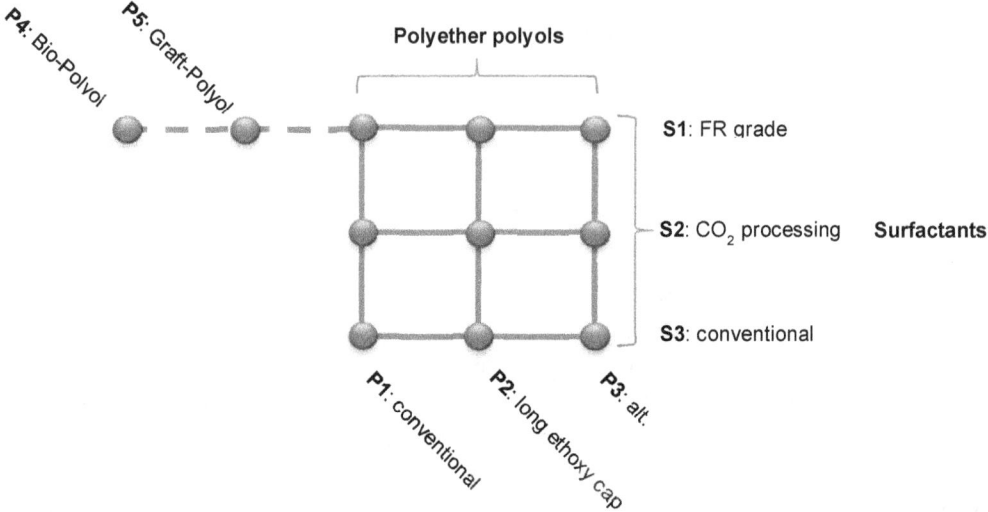

Figure 18. Design of experimental parameter space for Iteration 1.

The hydroxyl content for the polyols used in Iteration 1, expressed as *OH number* in Table 1, ranged between (31 to 57) mg KOH·g^{-1}. The TDI index (percentage ratio between the actual amount of TDI used in a formulation and the theoretical stoichiometric amount of TDI required to react with any reactive additive, *e.g.*, water and polyols) was kept constant at a value of 105 for all formulations by controlling the content of TDI, polyol, water, and reactive catalysts. The formulations were also designed to minimize differences in density (by tuning the water content) and air permeability (by tuning the catalyst content).

Nevertheless, significant differences in foam density and air permeability, observed among the foams of Iteration 1 (Table 4), were likely to influence smoldering and therefore the results in Iteration 1. For these reasons, the effect of secondary parameters deserved closer scrutiny. The secondary parameters investigated were foam density, air permeability, amount of tin catalyst, relative humidity during testing (R.H.-Testing), and air flow above the enclosure during testing. All these parameters may potentially affect smoldering.[4-11] These parameters were monitored and correlated to the final mass loss measured in 78 smolder box tests of Iteration 1.

Table 4 lists the range of values tested for these parameters along with the coefficients of correlation of the values of the secondary parameters to smoldering (assessed in the box test as the average mass loss measured at a heater target temperature of 340 °C and 360 °C). All of the coefficients were found to be very small. This demonstrates that the secondary parameters do not have a controlling effect on the outcome of the box tests.

Table 4. Influence of the secondary parameters on smoldering measured in the box test (78 tests).

Parameter	Unit	Min.	Max.	Avg	StDev	$R^{2\dagger}$
Foam Density	kg·m⁻³	27.5	30.8	29.1	1.0	< 0.01
Air Permeability	m·min⁻¹	45.9	90.9	80.2	29.1	0.02
Tin Catalyst Content	php[V]	0.12	0.21	§	§	0.02
R.H.-Testing	%	48	65	58	4	< 0.01
Air Flow above Enclosure	m·min⁻¹	0.07	0.12	0.08	0.02	0.05

† Coefficient of determination of a linear least square fit to mass loss after 100 min of test time.
§ The uncertainty of the dosing unit was not determined by the foam manufacturer.

3.1.2. Results for Smoldering in the Box Tests

The box test was designed to determine a minimum temperature required to initiate PUF smoldering and quantify smolder intensity. The minimum temperature required to initiate smoldering could be a potentially valuable property to describe the smoldering propensity of a given type of material. However, it is unclear if a distinct temperature of smolder initiation exists and, if so, how it would depend on foam properties and test conditions. In order to find the range of temperature at which smoldering is initiated in this particular setup, all foam samples were measured with heater set point temperatures of 320 °C, 330 °C, 340 °C, and 360 °C. Such a characterization required a large number of replicates for good statistical certainty on the minimum heater temperature required to assure self-sustained smoldering. Considering the number of formulations investigated, it was too time-consuming and unrealistic (due to the limited amount of foam available) to do more than replicate tests at each selected set temperature. It was decided that since the onset temperature of a formulation was unknown *a-priori,* there would be more information obtained by measurements at four temperatures with limited replicates than by measurements at one temperature with more replicates, but no information on the onset temperature.

Table 5 reports the mass loss measured in the box test for all formulations of Iteration 1 at the different set point temperatures. A total of 90 box tests were performed during Iteration 1. For each formulation, two or three tests were performed at 320 °C, one or two tests at 330 °C, two tests at 340 °C, and one or two tests at 360 °C. As previously mentioned, the number of replicates was limited by time and by foam-availability constraints.

Formulation A1R and B2 in Table 5 were repetitions of formulations A1 and A11, respectively.

Sustained smoldering[VI] was generally not observed at 320 °C except for A11 (in one test out of two) and A1R (single test only). At 330 °C, approximately 60 % of the samples were smoldering

---

[V] Parts (by mass) per hundred parts of polyol for a specific component in a foam formulation (*e.g.,* 10 php of X means that 10 g of component X are used in combination with 100 g of polyol). For clarity and convenience, php is used in place of mass fraction in foam formulations (a variation in the amount of a single additive in a formulation causes a variation in the mass fraction of all additives).
[VI] A mass loss above 2 % was observed in foams with sustained smoldering.

and at 340 °C and 360 °C, all of the samples smoldered (see Figure 19). Preliminary tests, run on a set of five commercial foams, gave similar results: a heater target temperature of 330 °C was sufficient to initiate self-sustained smoldering for four out of five samples.

In Table 6, the average percent mass loss and onset temperature for sustained smoldering are shown together with the polyol/surfactant type, density and air permeability. There is no significant correlation between density and permeability (secondary parameter) and onset smoldering temperature. Formulations based on polyol P3 always show an onset temperature between 320 °C and 330 °C, whereas, for all other polyols, the onset temperature appears to fall into a wider interval between 320 °C and 340 °C. As discussed below, the relatively low onset-temperature observed for P1 based formulations do not appear to significantly affect smoldering intensity, measured as mass loss in the box test.

Table 5. Mass loss measured in the box test at heater set-point-temperatures of 320 °C, 330 °C, 340 °C and 360 °C (90 tests).

Formulation	Test #	Heater at 320 °C			Heater at 330 °C			Heater at 340 °C			Heater at 360 °C		
		Mass loss (%)	Avg (%)	Dev (%)	Mass loss (%)	Avg (%)	Dev (%)	Mass loss (%)	Avg (%)	Dev (%)	Mass loss (%)	Avg (%)	Dev (%)
A1 (P1-S1)	1	1.0§	1.1	0.1	13.8	13.8	-	10.1	11.5	1.4	9.7	9.7	-
	2	1.2§			-			12.9			-		
A1R (P1-S1)	1	3.3	3.3	-	1.8§	2.7	0.9	5.6	5.5	0.1	7.9	7.9	-
	2	-			3.5			5.4			-		
A2 (P1-S2)	1	0.8§	0.9	0.1	1.4§	1.3	0.2	5.4	5.5	0.1	8.0	8.0	-
	2	1.0§			1.1§			5.5			-		
	3	0.8§			-			-			-		
A3 (P1-S3)	1	1.0§	1.0	<0.1	6.2	9.0	2.8	5.4	6.3	0.9	10.0	10.0	-
	2	1.0§			11.8			7.1			-		
A4 (P1/P4-S1)	1	1.0§	1.2	0.2	0.8§	0.7	0.2	6.6	6.2	0.4	4.3	4.3	-
	2	1.3§			0.4§			5.8			-		
	3	-			0.9§			-			-		
A5 (P1/P5-S1)	1	1.0§	1.0	0.1	1.1§	1.1	0.1	3.9	4.4	0.5	6.7	6.7	-
	2	0.9§			1.0§			4.9			-		
A6 (P2-S1)	1	1.1§	1.1	<0.1	4.2	5.2	1.0	7.0	6.6	0.5	7.2	7.2	-
	2	1.1§			6.2			6.1			-		
A7 (P2-S2)	1	1.2§	1.2	0.1	5.6	5.7	0.1	7.2	7.2	<0.1	8.4	8.4	-
	2	1.1§			5.7			7.2			-		
A8 (P2-S3)	1	0.8§	0.9	0.1	1.2§	1.2	<0.1	8.0	8.0	<0.1	4.7	4.7	-
	2	0.9§			1.2§			8.0			-		
A9 (P3-S1)	1	1.1§	1.0	0.1	6.8	8.6	1.8	7.7	7.8	0.1	11.4	11.4	-
	2	0.9§			10.3			7.8			-		
A10 (P3-S2)	1	1.0§	1.0	-	6.4	6.5	0.1	7.3	7.7	0.4	8.2	9.5	1.3
	2	-			6.6			8.1			10.8		
A11 (P3-S3)	1	6.1	3.5	2.6	7.2	7.7	0.5	8.5	7.7	0.8	6.9	6.9	-
	2	0.9§			8.1			6.9			-		
B2 (P3-S3)	1	0.1§	0.1	-	6.7	5.8	1.3	7.0	6.9	0.1	8.3	8.3	-
	2	-			4.8			6.8			-		

§Non-smoldering sample. The mass loss in non-smoldering samples is less than 2 %.

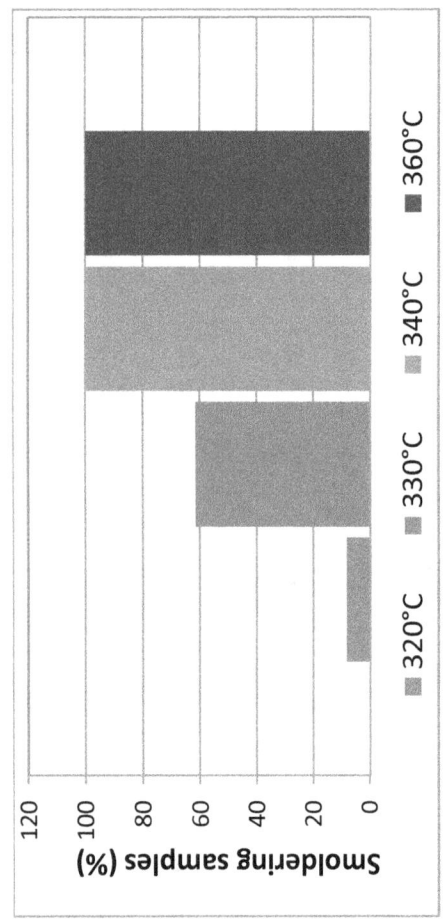

Figure 19. Percentage of samples from Table 6 showing self-sustained smoldering at various set-point temperatures of the heating element.

Table 6. Onset temperature for self-sustained smoldering, mass loss and main properties of formulations of Iteration 1. †Average mass loss in the box test for a heater set temperature of 340 °C and 360 °C (three replicates).

Formulation	Polyol-Surfactant	Density (kg·m⁻³)	Permeability (m·min⁻¹)	Onset temperature for self-sustained smoldering (°C)	AVG mass loss at 340 °C+360 °C ($ML_{340+360}$)† (%)
A1	P1-S1	30.1±0.8	77.2±2.3	320 to 330	10.9±1.7
A1R	P1-S1	28.4±0.4	45.9±7.7	320 to 330	6.3±1.4
A2	P1-S2	30.8±0.4	87.5±15.2	330 to 340	6.3±1.5
A3	P1-S3	30.1±0.9	70.9±8.6	320 to 330	7.5±2.3
A4	P1/P4-S1	29.1±0.6	89.2±3.2	330 to 340	5.6±1.2
A5	P1/P5-S1	28.7±0.2	76.3±10.9	330 to 340	5.2±1.4
A6	P2-S1	28.8±0.5	81.8±11.8	320 to 330	6.8±0.6
A7	P2-S2	29.4±0.8	90.9±9.8	320 to 330	7.6±0.7
A8	P2-S3	29.1±0.3	81.8±5.7	330 to 340	6.9±1.9
A9	P3-S1	27.7±0.3	90.4±5.2	320 to 330	9.0±2.1
A10	P3-S2	27.5±0.3	82.9±3.4	320 to 330	8.6±1.5
A11	P3-S3	28.8±0.2	77.7±8.6	320 to 330	7.4±0.9
B2	P3-S3	30.3±0.6	89.8±8.3	320 to 330	7.4±0.8

Figure 20 summarizes the mass loss *vs.* set point temperature of the heater in the box test. The value ranges (shown as uncertainty bars) are equal to the mean deviation.[VII] Data points that do not show an uncertainty bar are single measurements. The maximum mass loss was observed for sample A1 (13.8 %) at 330 °C; the mass loss measured for the same sample A1 at 340 °C and 360 °C was lower but there are not enough data available to conclude if this variation in mass loss between 330 °C and (340 or 360) °C is statistically significant.

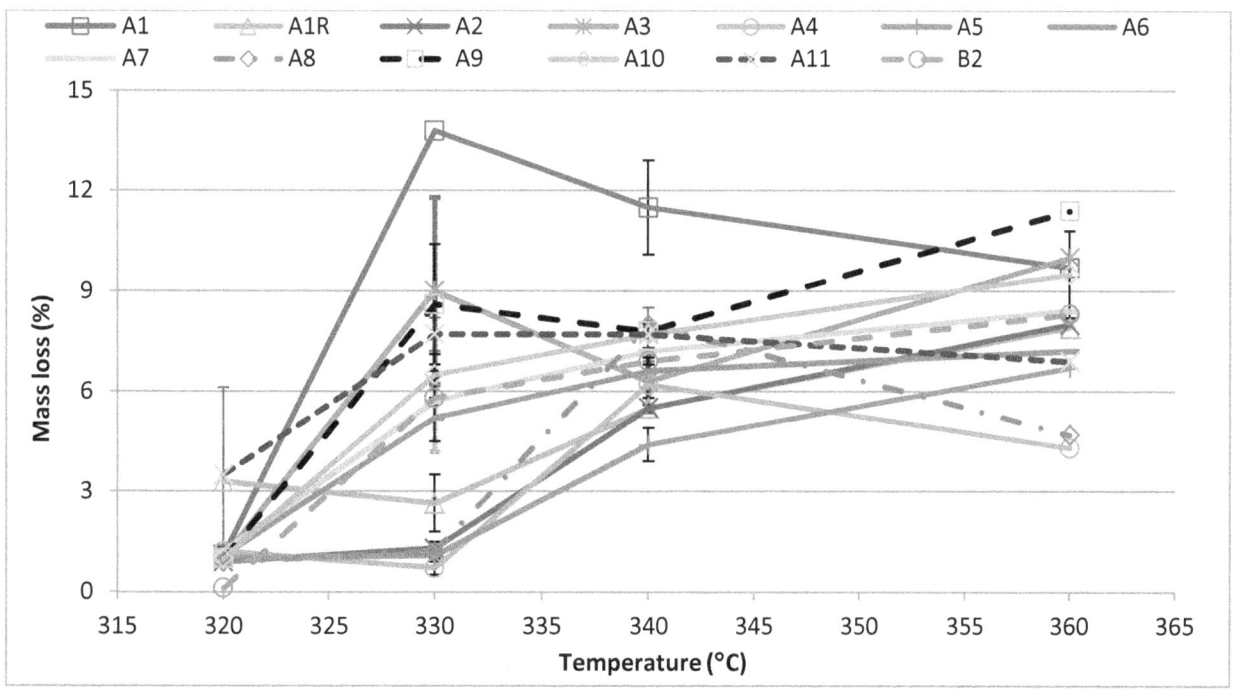

Figure 20. Mass loss *vs.* set point temperature of the heater in the box test using data from Table 5. The uncertainty bars shown here are equal to the average of the absolute deviations of data points from their mean (Dev). Data points that do not show an uncertainty bar are single measurements.

Figure 21 shows the average mass loss *vs.* set point temperature of the heater in the box test calculated for all samples in Table 6. Uncertainty bars shown here represent the range of the values measured during the test. This graph appears to be a sigmoid curve with an "S" shape. The mass loss is negligible below the onset temperature and then increases asymptotically with temperature. The high standard deviation observed at 330 °C is due to the fact that 330 °C is close to the smoldering onset temperature; about 60 % of the samples showed sustained smoldering at this temperature. The average difference in mass loss between 340 °C and 360 °C is about 1 % and is not statistically significant.

[VII] The mean deviation for the mass loss was determined as follows. The mass loss was determined in two replicate tests for each of nine PUF formulations. The difference between each pair of values was calculated and divided by two (deviation). The mean deviation, obtained as the average of the nine deviations, was equal to about 8 % of the mean mass loss.

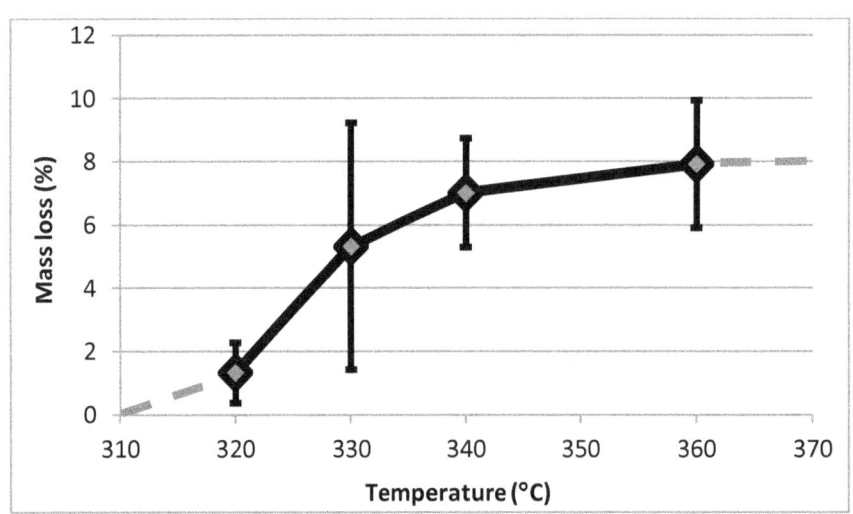

Figure 21. Average mass loss *vs.* set point temperature of the heater in the box test calculated for the samples in Figure 20 (uncertainty bars shown here are equal to one standard deviation).

Since only two tests at 340 °C and one test at 360 °C were conducted, an average of the mass loss measured at 340 °C and 360 °C (temperatures at which all samples demonstrated smoldering) was calculated for a more reliable assessment of the smoldering intensity (Table 6). The average mass losses at 340 °C and 360 °C ($ML_{340+360}$) for all the formulations are shown in Figure 22. Formulation A1 has the highest $ML_{340+360}$ value. Figure 23 reports the values of $ML_{340+360}$ grouped by surfactant or polyol type. For example, $ML_{340+360}$ shown for S1 is an average of the $ML_{340+360}$ measured for A1, A6, and A9 and $ML_{340+360}$ shown for P2 is an average of the $ML_{340+360}$ measured for A6, A7, and A8. The analysis of variance indicates that there is no significant difference between the mean values of groups P1, P2 and P3 or groups S1, S2 and S3 (Figure 23). This means that in terms of mass loss there is no systematic effect of the polyol or the surfactant on smoldering.

The repeatability of the foaming process was verified by manufacturing replicates of the formulations. For example, A1R and B2 are chemical replicates (same formulation but foamed at different temperature and relative humidity, see Appendix 1 for details on atmospheric conditions, processing parameters and foam properties) of A1 and A11, respectively. Polyol P3 used in formulation A1 and A1R appeared to be insensitive to variations in atmospheric conditions and smoldering performance was unaffected. However, there was a significant variation between the two formulations based on polyol P1 (A1R and B2) in terms of average mass loss ($ML_{340+360}$ = 10.9±1.7 % for A1, and $ML_{340+360}$ = 6.3±1.4 % for A1R), detailed in Figure 24.[VIII] Noticeably, even though A1 and A1R are identical formulations, they show the highest and lowest $ML_{340+360}$, respectively, in the conventional PUF formulation (*i.e.*, not including graft-polyol P5 or bio-polyol P4) (Table 6). As also reported in Table 6, the air permeabilities and densities were largely different for A1 and A1R. These results indicate that atmospheric conditions during manufacturing may cause substantial variation of foam morphology that overshadows any effect on smoldering created by the chemical differences associated with the three primary polyols and/or three surfactants selected for this study.

[VIII] The relatively higher sensitivity of polyol P1 to variations in atmospheric conditions was confirmed by the foamer on the basis of its multi-year experience.

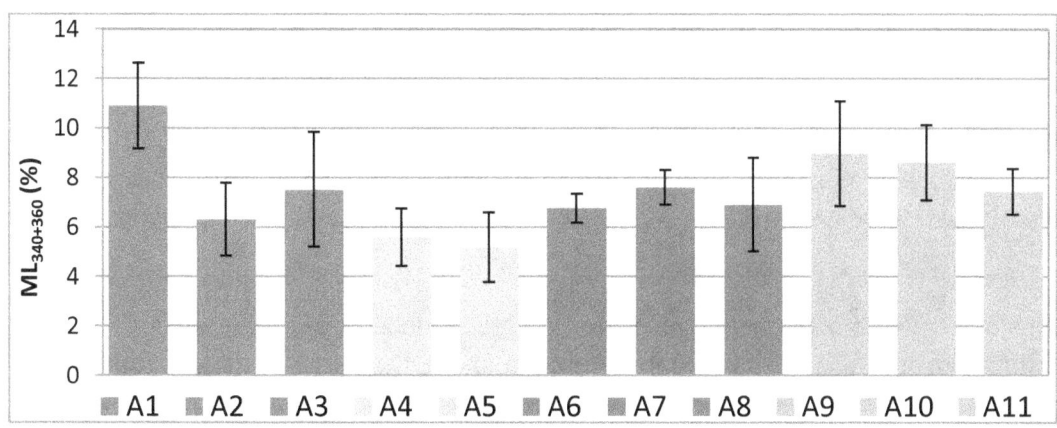

Figure 22. Average mass loss at 340 °C and 360 °C ($ML_{340+360}$) in the box test for the formulations of Table 6 (uncertainty bars equal to one standard deviation). The $ML_{340+360}$ for the replicates of formulation A1 and A11 (A1R and B2, respectively) are shown in Figure 24.

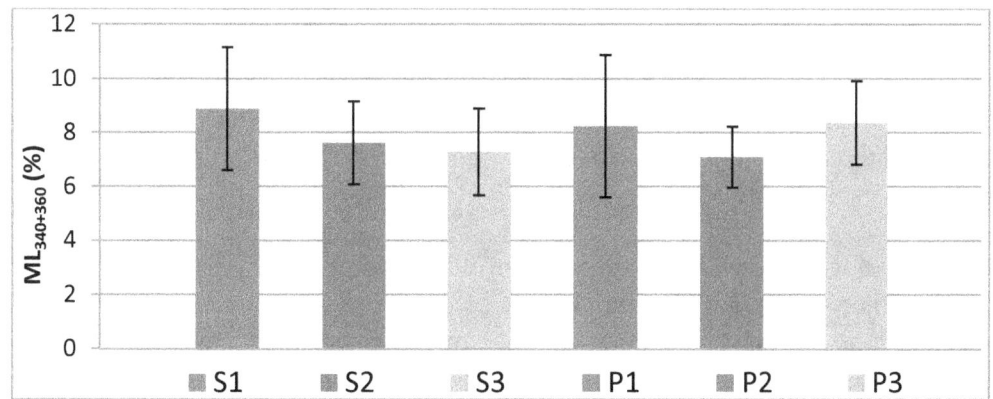

Figure 23. Average mass loss at 340 °C and 360 °C ($ML_{340+360}$) in the box test grouped by surfactant or polyol type. For example, $ML_{340+360}$ shown for S1 is an average of the $ML_{340+360}$ measured for A1, A6 and A9 (uncertainty bars equal to one standard deviation).

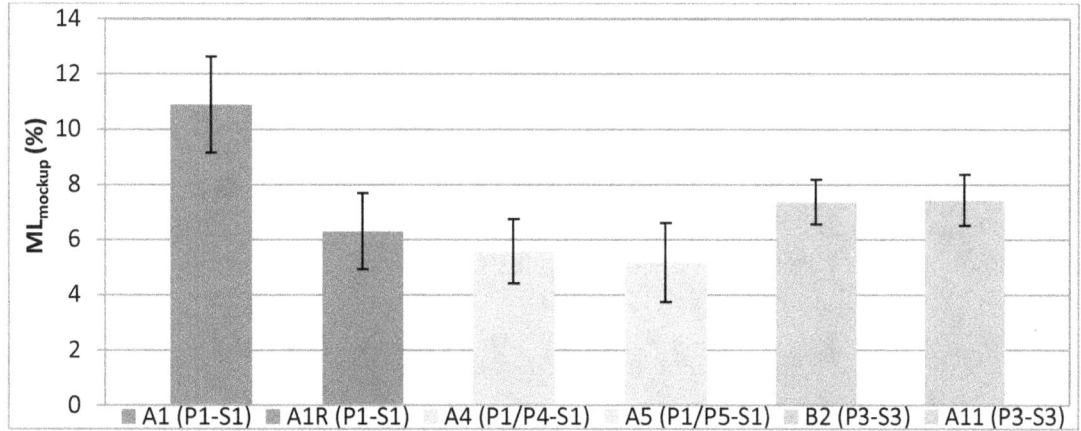

Figure 24. Average mass loss at 340 °C and 360 °C ($ML_{340+360}$) in the box test for formulations A1, A1R, A4, A5, B2 and A11 (uncertainty bars equal to one standard deviation).

The non-conventional formulations, A4 and A5, are also compared in Figure 22. Replacing 25 % of polyol P1 with a bio-polyol in A4 or graft-polyol in A5 produced a large reduction in $ML_{340+360}$ as compared to A1 but no significant reduction as compared to A1R. The effect of adding P4 and P5 to the formulation on smoldering might be explained by a change of foam morphology rather than foam chemistry. Therefore, the morphology of formulations A1, A1R, A4 and A5 was investigated by means of a confocal microscope (Zeiss LSM 510). An optical section with an identical thickness of 1 mm was collected for all samples for comparison of cell structure.

The micrographs of Figure 25 show heterogeneities in A1R produced by large clusters of residual membranes (likely fully closed cells) in the right and bottom area and a generally more open and homogeneous cell structure for A1. This is consistent with the significantly different measured values of air permeability and smoldering propensity for A1 and A1R.

Since A1 and A5 have similar air permeabilities, it was expected that the smoldering behavior would also be similar; however, the smoldering intensity of the A5 foam was significantly smaller than that of the A1 foam. The air permeability of the A4 foam was greater than the A1 foam, but A4 showed a lower smoldering propensity than the A1 foam (Table 6). Figure 26 shows that the cell structures of formulations A4 and A5 were very similar. There were a very limited number of residual membranes and the average cell size was larger than in A1. Both of these characteristics caused a reduction in surface area of the foam. These data suggest that surface area also plays a key role in smoldering.

The idea that smoldering performance is dominated by morphology and not chemistry for the set of formulations tested here is confirmed by the mockup tests. Tests were conducted with three mockups each for formulations A1, A1R, A8, and A11. The mass loss values measured by the box test ($ML_{340+360}$) and the mockup test (ML_{mockup}) for these formulations are compared in Table 7. The same data are also plotted in Figure 27. The observed morphologic differences between the chemically identical A1 and A1R formulations result in an even larger difference in smoldering intensity in the mock-up test as compared to the box test; the average $ML_{340+360}$ increased from 6.3 % for A1R to 10.9 % for A1 (the ratio between $ML_{340+360}$ for A1 and $ML_{340+360}$ for A1R is about 1.7), and the average ML_{mockup} increased from 0.5 % for A1R up to 30.5 % for A1 (the ratio between ML_{mockup} for A1 and ML_{mockup} for A1R is 61).

The results in Figure 27 show that the box test is not as sensitive as the mockup test to morphological variations. Formulations A11 and A8 have a comparable mass loss in the box test to formulation A1R but drastically different mass loss in the mockup test. This is due to the fact that, as discussed in Section 3.2.8, there is a threshold value for air permeability ($\Phi \approx 45$ m·min^{-1}), below which no smoldering is observed in the mockup test. A detailed comparison of data outputs from the two tests is discussed in Section 3.2.3.

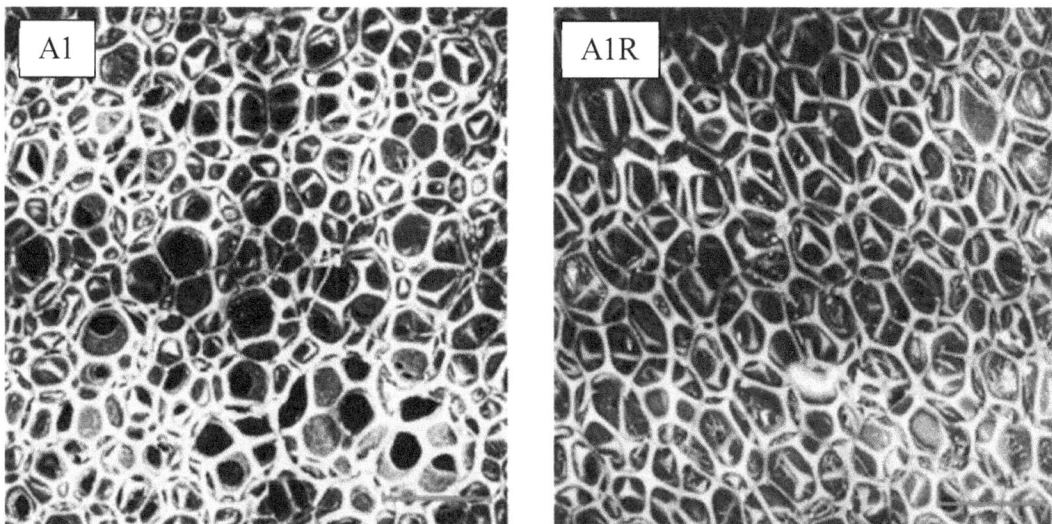

Figure 25. Optical micrographs for formulations A1 (on the left) with an air permeability of (77.2 ± 2.3) m·min^{-1} and its replicate formulation A1R (on the right) with an air permeability of (45.9 ± 7.7) m·min^{-1}. Bar size in the right bottom corner is 1 mm.

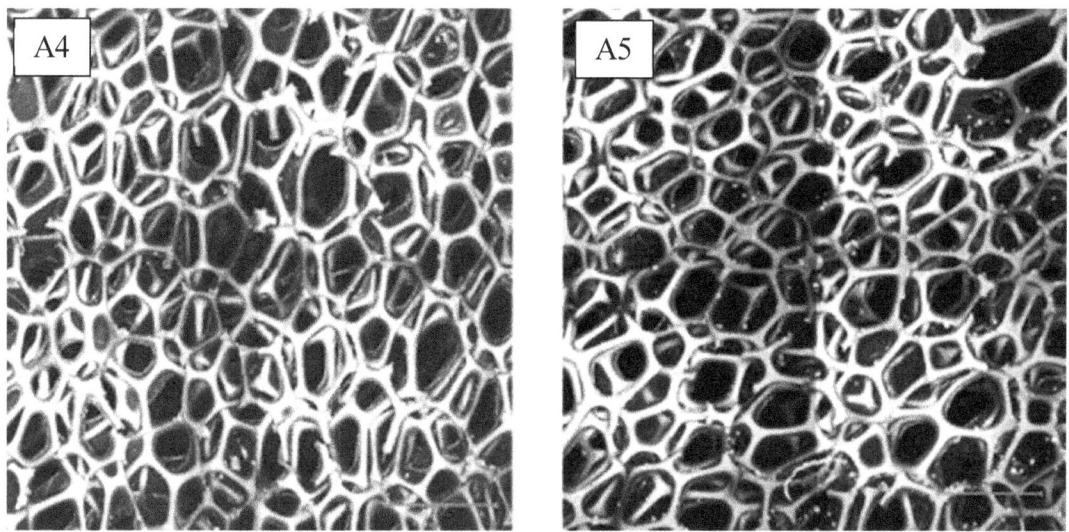

Figure 26. Optical micrographs for: A4 (bio-polyol formulation) with an air permeability of (89.2 ± 3.2) m·min^{-1} (on the left); and A5 (graft-polyol formulation) with an air permeability of (76.3 ± 10.9) m·min^{-1} (on the right). Bar size in right bottom corner is 1 mm.

Table 7. Mass losses measured by the box test (ML$_{340+360}$) and the mockup test (ML$_{mockup}$) are compared for 4 formulations in Iteration 1 (uncertainty is shown as one standard deviation calculated over at least 3 replicates).

Formulation	ML$_{mockup}$ (%)	ML$_{340+360}$ (%)	Air Permeability, Φ (m·min^{-1})
A1	30.5±12.1	10.9±1.7	77.2±2.3
A1R	0.5±0.2	6.3±1.4	45.9±7.7
A8	32.0±2.3	6.9±1.9	81.8±5.7
A11	29.3±6.5	7.4±0.9	77.7±8.6

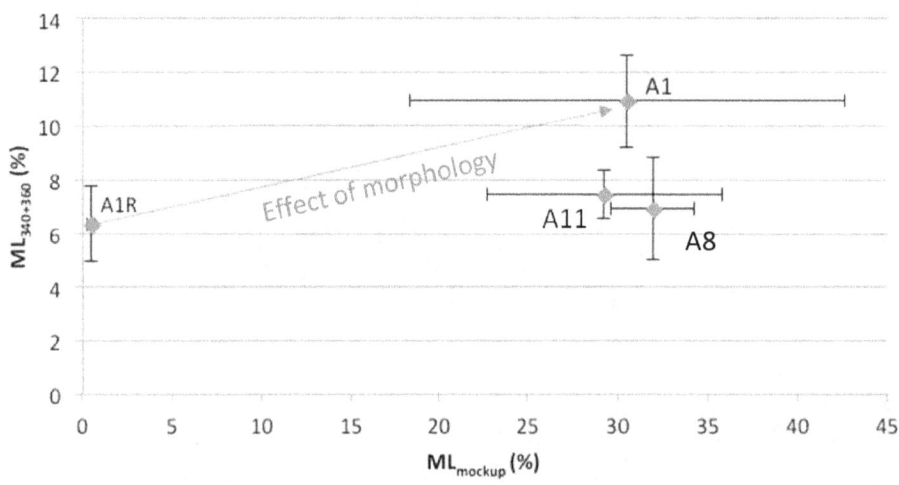

Figure 27. Mass loss in the box test ($ML_{340+360}$) *vs.* mass loss in the mockup test (ML_{mockup}) for four formulations of Iteration 1 (uncertainty bars equal to one standard deviation).

3.1.3. Results for Flame Spread Tests

Flame spread tests were conducted to examine the role of surfactants on the burning behavior of the foams and evaluate the possible effects on the CPSC proposed standard. All formulations from Iteration 1 were tested, each in three replicates. Typically, the flame spread over the surface of the sample, and then the foam started to collapse and form a liquid pool. The results demonstrated that the surfactant had a strong impact on the burning behavior of the pool. Surfactants S1 and S2 are designed to decrease the flammability of PUF and S3 is a conventional non-flame-retardant surfactant. Foams containing S1 and S2 showed reduced mass loss rate (MLR) for all three polyols as compared to S3 (Figure 28). Surfactants with flame-retardant action may suppress flaming over the liquid produced by thermal degradation of the foam. This suggests that surfactant type can also have an impact on the open-flame test according to the CPSC proposed standard, where a pool fire may occur.

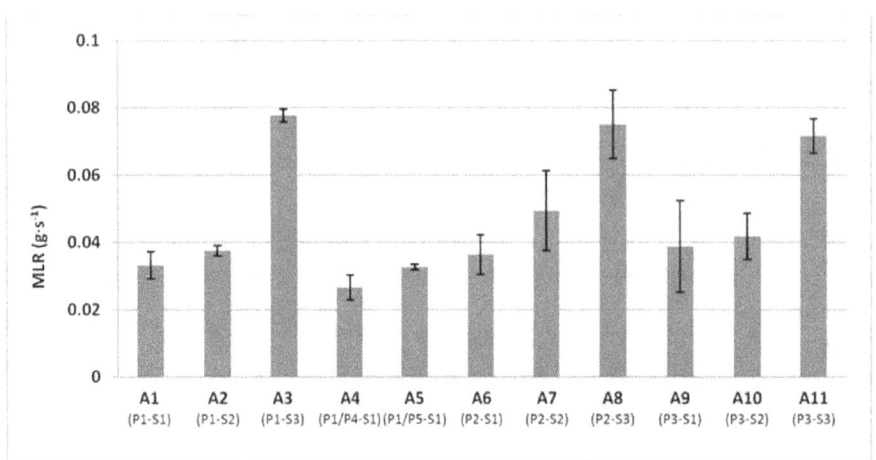

Figure 28. MLR measured for Iteration 1 formulations. There is a significant reduction of MLR for surfactants S1 and S2. Uncertainty shown as one standard deviation calculated over at least 3 replicates.

28

However, in terms of average flame spread rate (FSR) a fairly low variation for different foam formulations was observed (Figure 29). The formation of pool fires with S3 led to a small increase in FSR with polyols P1 and P2, but not with P3. These results do not show a significant systematic effect of a specific polyol on mass loss or flame spread.

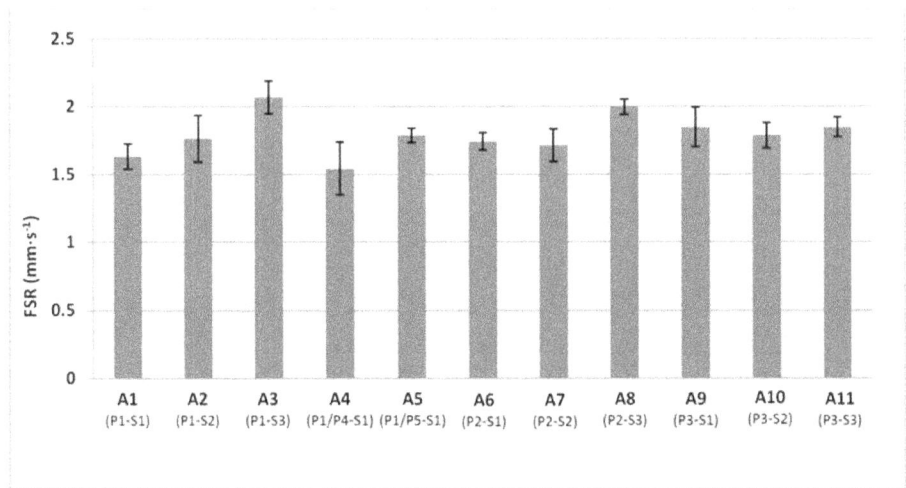

Figure 29. FSR measured for Iteration 1 formulations. Uncertainty is shown as one standard deviation calculated over at least 3 replicates.

3.1.4. Conclusions of Iteration 1

In Iteration 1, three types of commercially available polyols and surfactants were selected in order to investigate the effect of chemical composition on smoldering. Formulations with all possible combinations of three different polyols with three different surfactants were manufactured in order to investigate whether there is a systematic effect of either one of the polyols or surfactants. In terms of smoldering (measured as mass loss in the box test), there is no systematic effect on smoldering for any of the polyols or surfactants (within the parameters chosen for this project). The results demonstrated that the variations in foam morphology due to variation in manufacturing conditions (*e.g.*, temperature and humidity) may override any potential effect of the chemical composition. In terms of the open-flame test, surfactants with a physical flame-retardant action may prevent flame propagation over the liquid pool and decrease the mass loss rate.

3.2. Iteration 2: Impact of Processing Parameters

The purpose of Iteration 2 was to determine if the processing parameters (within the parameters chosen for this project) impact PUF smoldering and open-flame performance. Typical processing parameters with a strong impact on foam morphology are water content, tin catalyst content and head pressure. They are routinely adjusted during PUF manufacturing to compensate for morphological variations caused by climatic conditions (*i.e.*, variations in atmospheric temperature, humidity and pressure) and deliver a PUF with consistent specifications throughout the year. Water controls the blowing action by reacting with TDI and releasing CO_2, the tin catalyst controls the rate of the polyol/TDI reaction, and the head pressure controls nucleation and cell growth at the exit of the mixing head.[21]

A 2^k3 factorial experimental design (two levels for three factors) with high and low levels of water, tin catalyst, and mixing head pressure was chosen in order to produce flexible foam samples with a range of densities (28.4 kg·m^{-3} to 34.3 kg·m^{-3}) and air permeabilities (3.1 m·min^{-1} to 74.2 m·min^{-1}). Figure 30 shows a schematic diagram with the experimental design and effect of each factor on foaming. The choice of the experimental-design factors was limited to the three parameters expected to lead to the most significant changes in foam morphology. This does not exclude the possibility that other parameters could also significantly affect smoldering.

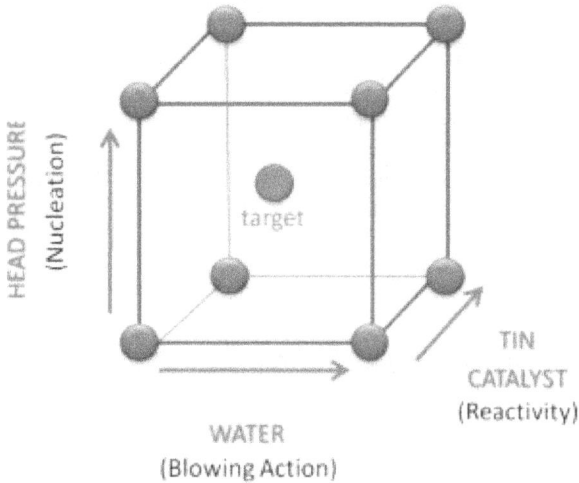

Figure 30. Schematic diagram showing the 2^k3 experimental design with high and low settings of the amount of water, amount of tin catalyst and mixing head pressure.

Iteration 1 showed that there was no systematic effect on smoldering of any specific polyol or surfactant, so the selection of the formulation for Iteration 2 was not critical. Originally, formulation A8 (P2-S3) was selected, because it is a standard PUF based on a conventional surfactant without any flame retardant action and a widely used commercial polyol. However, a different formulation, A11 (P3-S3) was ultimately used in combination with a processing aid, because formulation B1 (replicate of A8 during Iteration 2) yielded foams with poor quality due to splitting (cracks in the bun) because it was manufactured in a different climate (see Appendix 1). The average manufacturing plant temperature and relative humidity decreased from (25.0 ± 1.8) °C and (59.3 ± 10.7) % during Iteration 1, to (21.6 ± 0.7) °C and (28.1 ± 4.0) % during Iteration 2, respectively. According to the foamer and to our results from Iteration 1, polyol P1 is more sensitive to temperature/humidity variations than polyol P3. As already mentioned, the choice of the polyol is not critical so polyol P1 was replaced by polyol P3. However, the use of polyol P3 did not solve the splitting problem as formulation B2 gave rise to the same foam cracks observed in B1 foam.

At this point, due to the inability to control temperature and humidity in the pilot plant the only possible solution was to vary the formulation. The foamer suggested the addition of 0.15 php[IX] of a processing aid (Table 3), which is a mixture of emulsifiers based on a fatty ester. This processing aid was used to decrease the viscosity and enhanced flow of the foaming material in the tube connecting the mixing head with the foaming box. This was sufficient to prevent foam splitting in formulation B3.

With a suitable formulation agreed upon, the next step was to investigate the effect of processing parameters on smoldering. Ideally, the processing parameters should be varied over a wide range, but this was not possible due to limited foam stability at the extreme processing conditions. A series of preliminary experiments was performed to identify the processing window, *i.e.*, the maximum range of variability for water content, tin catalyst content and mixing head pressure (referred to hereafter as "pressure") that ensures successful foaming.

Formulation B3 was the first of Iteration 2 that yielded good quality foams and was not affected by splitting. In the B3 foam, the tin catalyst content was 0.2 php, the water content 2.95 php and pressure 34.5 kPa (5 psi) (see Appendix 1.) Initially, 0.2 php was selected as the high level for the tin catalyst, 2.95 php the high level for water and 34.5 kPa as the low level for pressure. When the water level was decreased from 2.95 php to 2.7 php (low level for water content) there was excessive shrinkage and the high level of tin was decreased to 0.19 php. For this reason formulation B3 was superseded and formulation B4 was used instead in the 2^k3 experimental design. Similar iterations were carried out for identifying feasible low/high levels for pressure and tin content.

The selected high and low levels for the three processing parameters are shown in Table 8. Outside of this processing window foaming was not reproducible or yielded poor quality foams due to foam collapse, shrinking, or splitting.

Table 8. Summary of the selected high and low levels used for tin catalyst content, water content and head pressure.

Processing parameter	Low Level	High level
Tin Catalyst (php)	0.16	0.19
Water (php)	2.70	2.95
Head Pressure (kPa)	34.5	55.2

The eight formulations of Iteration 2 used for the full factorial experimental design are provided in Figure 31 and Table 9. The average air permeability and density for the eight formulations used in the experimental design and three additional formulations from Iteration 2 (B2, B3 and B6) were calculated over at least four buns (Table 10). For convenience, the smoldering performance is also reported in Table 10 and discussed in the following sections. The average permeability calculated for these samples from Iteration 2 was (31 ± 28) m·min⁻¹ while the average permeability of the samples from Iteration 1 was (80 ± 12) m·min⁻¹. This reduction in

---

[IX] Parts per hundred polyol (php) is the mass of a component in a foam recipe defined relative to the largest mass component of the formulation; the polyol.

permeability was thought to be mainly due to a variation in climatic conditions, but the processing aid might also have a small effect on air permeability (see Iteration 3).

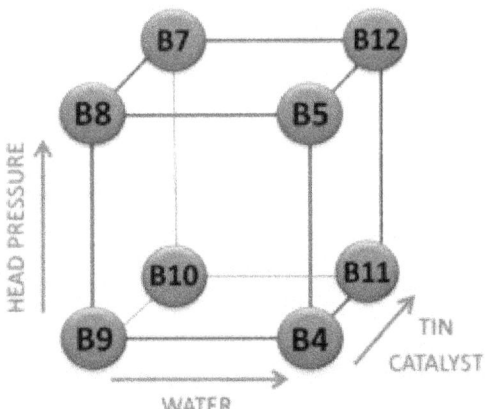

Figure 31. Schematic diagram showing the 2^k3 experimental design with the corresponding formulations.

Table 9. Formulations of Iteration 2 and corresponding levels for the full factorial experimental design.

	B4	B5	B7	B8	B9	B10	B11	B12
Water	High	High	Low	Low	Low	Low	High	High
Tin catalyst	Low	Low	High	Low	Low	High	High	High
Pressure	Low	High	High	High	Low	Low	Low	High

Table 10. Density (ρ), air permeability (Φ), mass loss in the mockup test (ML_{Mockup}) and mass loss in the box test ($ML_{340+360}$) for the foams of Iteration 2. The average values (Avg) ± one standard deviation (StDev) are reported.

	ρ (kg·m^{-3})	Φ (m·min^{-1})	$ML_{340+360}$ (%)	ML_{mockup} (%)
B2*	30.3±0.6	89.8±8.3	7.4±0.8	-
B3*	30.2±0.6	49.3±10.6	2.4±1.6	-
B4	30.1±0.6	74.2±8.4	6.8±0.5	8.1±3.4
B5	30.4±0.3	44.8±4.1	2.2±1.9	0.2±0.1
B6*	33.8±0.8	33.4±5.5	2.9±1.1	0.5±0.2
B7	31.6±0.4	4.7±1.7	0.8±0.2	0.5±0.6[†]
B8	33.6±0.5	16.6±4.1	1.4±0.5	0.7±0.2[†]
B9	34.3±0.6	59.7±3.4	5.1±2.3	15.2±5.6
B10	30.7±0.7	3.1±0.4	0.8±0.6	1.2±0.5[†]
B11	29.2±0.6	43.3±5.9	5.4±2.2	2.5±2.2
B12	28.4±0.6	3.4±3.3	0.8±0.2	0.3±0.3[†]

*This formulation is not part of the experimental design.

† Formulations that did not smolder in the box test at the highest heater set temperature (360 °C) did not smoldered in the mockup test either.

Three replicates of each of the formulations were smoldered in the box test and three replicates were smoldered in the mockup test. The values of average mass loss at a heater temperature of 340 °C or 360 °C ($ML_{340+360}$) and the average mass loss in the mockup test (ML_{mockup}) are also shown for convenience in Table 10 and discussed further in the following sections.

3.2.1. Bun-to-bun variability and in-bun variability

As discussed above, climatic conditions (temperature, humidity and atmospheric pressure) affect the cell morphology so a fine tuning in processing parameters (water content, catalyst content and head pressure) is required on a daily basis for a consistent foam quality. This implies that even for a given formulation, a bun-to-bun variability in the foam properties is likely. Significant variations in PUF properties are also observed in the same foam bun (in-bun variability) as a function of the specific location due to surface effects and pressure/temperature gradients.

For in-bun variability, multiple measurements of air permeability for each bun were performed by NIST to assess potential variations in permeability along the direction of pouring. Foam slices were cut perpendicular to the direction of pouring. Eight measurements per slice were performed. The foam buns were about 90 cm long. The results are shown in Figure 32 (Iteration 1) and Figure 33 (Iteration 2). Air permeability varied dramatically along the pouring direction. For formulations from Iteration 1, it generally reached a maximum in the center of the bun and decreased towards the pour start and pour end. For formulations from Iteration 2, there was no clear trend. The processing conditions for these foams were purposely selected on the edges of the processing window. They are relatively instable formulations with irregular cell structure and, consequently, irregular air permeability.

The bun-to-bun variability was assessed by calculating the standard deviation of the air permeability measured by the foamer for four or more different buns of the same formulation. The bun-to-bun variability calculated for the same foams of Figure 32 and Figure 33 are reported in Table 11 and compared to the in-bun variability.

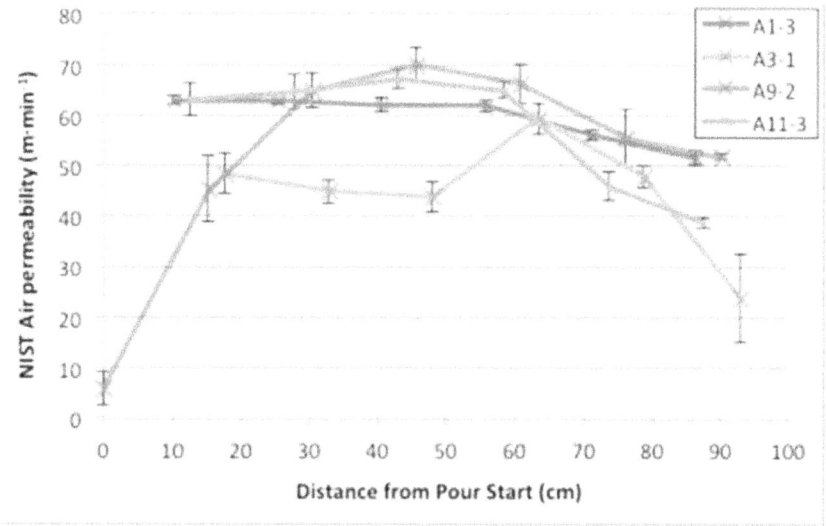

Figure 32. Air permeability *vs.* distance from the pour start for formulations of Iteration 1 (uncertainty bars are equal to one StDev over at least 3 replicates).

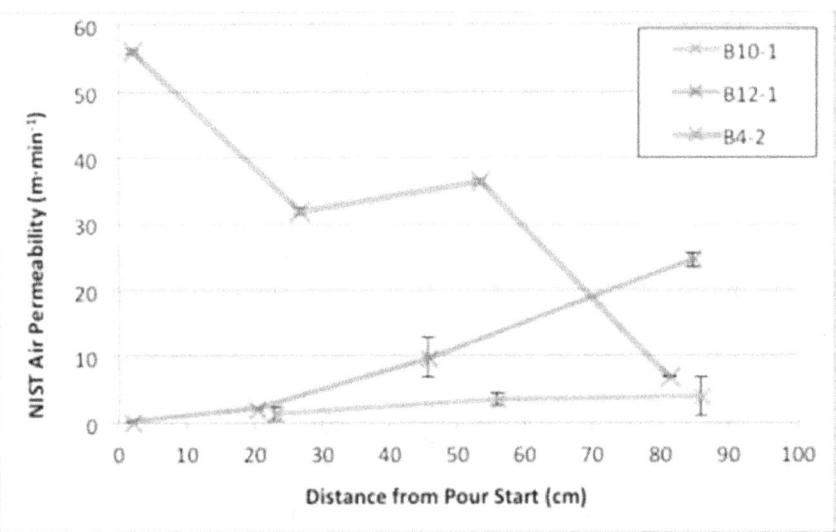

Figure 33. Air permeability *vs.* distance from the pour start for formulations of Iteration 2 (uncertainty bars are equal to one StDev over at least 3 replicates).

The in-bun variability in terms of relative standard deviation (StDev%) of the air permeability (Φ) is generally higher than the bun-to-bun variability. In other words, the properties of a foam sample are more affected by the position of the sample in the bun, rather than the specific bun used. This is true for the small buns used here, prepared in a pilot plant, but much lower in-bun variability is expected for large size buns from a production line.

Table 11. Comparison between in-bun variability and bun-to-bun variability in terms of air permeability (Φ). The average values (Avg), the standard deviation (StDev) and relative standard deviation (StDev%) are reported.

Formulation		In-Bun variability Permeability, Φ†			Bun-To-Bun variability Permeability, Φ‡		
	Bun #	Avg (m·min^{-1})	StDev (m·min^{-1})	StDev% (%)	Avg (m·min^{-1})	StDev (m·min^{-1})	StDev% (%)
A1	3	59.6	4.7	7.9	77.1	5.2	6.7
A3	1	44.8	11.7	26.0	70.8	8.6	12.1
A9	2	51.5	22.0	42.7	90.3	5.2	5.8
A11	3	57.6	12.0	20.9	77.7	8.6	11.1
B4	2	32.9	20.2	61.6	74.2	8.4	11.3
B10	1	2.9	1.4	47.5	3.1	0.4	12.9
B12	1	9.2	11.1	121.2	3.4	3.3	97.1

†Values of Φ measured in different locations of the same foam bun by NIST.
‡Values of Φ measured in different foam buns and same middle section of the bun by the foamer.

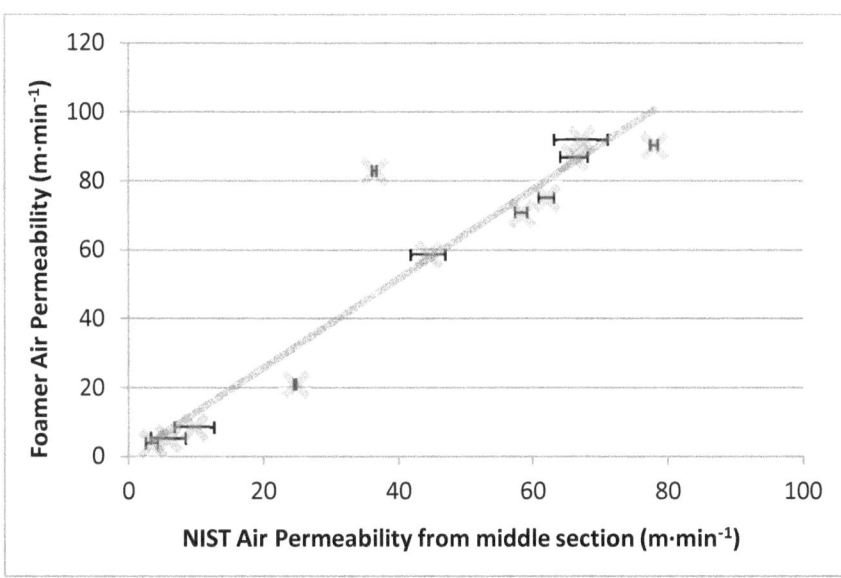

Figure 34. Correlation between the air permeability measured by the foamer in the pilot plant and air permeability measured by NIST collected from the middle section of the foam bun (30 cm middle section between pour-start and pour-end). Uncertainty bars for NIST data are equal to the StDev over at least 3 replicates. The foamer's data are based on a single measurement per bun. The red line is a least-squares regression fit to the data points.

The foam slices used for the air permeability measurements by the foamer and NIST were cut from planes that were orthogonal to each other (see Section 2.5). Due to anisotropy in the foam, cells often looked roughly like circles in the foamer's slices and ovals in the NIST slices. As a result of the anisotropy in the foam, the air permeability was measured to be 30 % higher in the foamer's samples than in the NIST samples (Figure 34). The NIST air permeability data reported here were calculated only from the middle section of the foam bun (30 cm middle section between pour-start and pour-end) to minimize in-bun variability. Similarly, all the foamer's data were collected from the middle section. Except for one outlier, there is a good correlation between the two sets of measurements. The values of permeability measured by the foamer were used to assess foam permeability in the remaining tests described below (unless otherwise specified). The permeability values appeared to be reliable for the middle section of the foam buns; however, outside of this region, the variation in permeability can be drastic.

3.2.2. Smoldering performance - Box test

The complete set of smoldering data from the box test for all formulations of Iteration 2 at a heater-set-point temperature of 320 °C, 330 °C, 340 °C, and 360 °C are reported in Table 12. A drastic decrease in smoldering was observed in Iteration 2 as compared to Iteration 1, likely due to the lower values of foam air permeability. The average value of $ML_{340+360}$ was 3.3 ± 2.5 % for the formulations of Iteration 2 and 7.3 ± 1.5 % for the formulations of Iteration 1 (Table 12). Foams from Iteration 2 were less smolder-prone than samples from Iteration 1. The percentage of smoldering samples was also lower in Iteration 2 as compared to Iteration 1 for a given temperature. Sustained smoldering in foams from Iteration 2 was never observed at 320 °C; at 330 °C about 35 % of the samples were smoldering; at 340 °C about 39 % were smoldering and even at the highest temperature (360 °C) only about 64 % of the samples were smoldering

(Figure 35). A higher heater set-point temperature would be required to fully investigate the dependence of smoldering *vs.* temperature in Iteration 2. Tests at a higher heater set-point were not conducted due to limited foam availability and the smoldering behavior previously observed in commercial foams.[X]

Table 12. Mass loss for formulation of Iteration 2 measured in the box test at a heater-set-point temperature of 320 °C (ML_{320}), 330 °C (ML_{330}), 340 °C (ML_{340}), 360 °C (ML_{360}) and the average of the mass loss measured at a temperature of 340 °C and 360 °C ($ML_{340+360}$).

		MASS LOSS (%)									
		ML_{320}		ML_{330}		ML_{340}		ML_{360}		$ML_{340+360}$	
	Test #	One Test	Avg (Dev)	One Test	Avg (Dev)	One Test	Avg (Dev)	One Test	Avg (Dev)	Avg	StDev
B2*	1	0.1§	0.1	6.7	5.8	7.0	6.9	8.3	8.3	7.4	0.8
	2	-		4.8	(1.3)	6.8	(0.1)	-	-		
B3*	1	0.7§	0.7	2.2§	1.2	2.0§	1.6	4.2	4.2	2.4	1.6
	2	-	-	0.2§	(1.0)	1.1§	(0.5)	-	-		
B4	1	0.9§	0.9	5.3	4.4	6.9	6.6	7.2	7.2	6.8	0.5
	2	-	-	3.5	(0.9)	6.2	(0.4)	-	-		
B5	1	0.9§	0.9	0.9§	0.9	1.3§	1.1	4.3	4.3	2.2	1.9
	2	-	-	0.9§	(0.0)	0.9§	(0.2)	-	-		
B6*	1	0.8§	0.8	0.8§	1.4	1.7§	2.4	3.9	3.9	2.9	1.1
	2	-	-	1.9§	(0.6)	3.1	(0.7)	-	-		
B7	1	0.5§	0.5	0.5§	0.6	0.6§	0.7	0.9§	0.9	0.8	0.2
	2	-	-	0.6§	(0.1)	0.8§	(0.1)	-	-		
B8	1	0.6§	0.6	0.9§	0.8	2.0§	1.6	1.2§	1.2	1.4	0.5
	2	-	-	0.6§	(0.2)	1.1§	(0.5)	-	-		
B9	1	0.5§	0.5	5.6	5.8	3.6	3.8	7.8	7.8	5.1	2.3
	2	-	-	5.9	(0.2)	4	(0.2)	-	-		
B10	1	-	-	0.7§	0.7	0.2§	0.6	1.4§	1.4	0.8	0.6
	2	-	-	0.7§	(0.0)	0.9§	(0.4)	-	-		
B11	1	0.7§	0.7	4.2	3.4	7.3	5.3	5.6	5.6	5.4	2.2
	2	-	-	2.3§	(0.7)	2.2§	(2.0)	-	-		
	3	-	-	3.6		6.3		-			
B12	1	0.9§	0.9	0.8§	0.7	0.7§	0.9	0.8§	0.8	0.8	0.2
	2	-	-	0.6§	(0.1)	1.0§	(0.2)	-	-		

*This formulation is not part of the experimental design.

†Average mass loss calculated at temperature heater of 340 °C and 360 °C.

‡ Average of the absolute deviations of data points from their mean.

§ Non-smoldering sample.

[X] Preliminary tests on commercial grade PUF showed that some formulations that did not smolder at 360 °C did also not smolder at 400 °C.

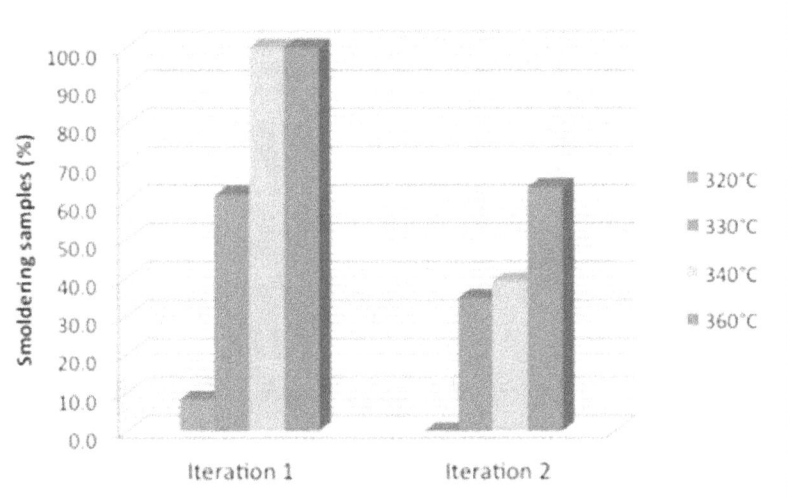

Figure 35. Percentage of samples showing self-sustained smoldering in the box test *vs.* set-point-temperatures of the heater for formulations of Iteration 1 and Iteration 2. The percentage of smoldering samples is reduced in Iteration 2 as compared to Iteration 1 for a given temperature.

Figure 36 shows the average mass loss *vs.* set point temperature of the heater in the box test calculated for samples of Iterations 1 and 2. The average mass loss is significantly lower in Iteration 2 as compared to Iteration 1 at all temperatures. The plot for Iteration 1 is a sigmoid curve with an "S" shape whereas the same plot for Iteration 2 is almost perfectly linear in the temperature range investigated. The different trend can be explained by a wider distribution in onset temperature: in Iteration 1, self-sustained smoldering was always observed at an heater set-point temperatures above 330 °C, but in Iteration 2 there were formulations with an onset temperature between 320 °C and 330 °C (B2, B4 and B9), formulations with an onset temperature above 360 °C (B7, B8, B10, B12), and a formulation (B11) with an onset spanning between 320 °C and 360 °C, likely due to foam heterogeneity.[XI]

As reported in in Table 10, the formulations with the high air permeability (B2, B4 and B9) showed a low onset temperature (between 320 °C and 330 °C), the formulations with intermediate air permeability (B3, B5, B11 and B6) showed an intermediate onset temperature (between 340 °C and 360 °C), and formulations with a low air permeability (B8, B7, B12 and B10) showed a high onset temperature (360 °C and above). These data suggest that the onset temperature for sustained smoldering in the box test depends significantly on air permeability. This might be related to the increase in oxygen supply generated by buoyant convection when the temperature of the heating element increases.

[XI] As discussed in Section 3.2.1, the processing conditions for Iteration 2 foams were purposely selected on the edges of the processing window. Consequently, they are relatively instable formulations with large variations in cell structure and air permeability.

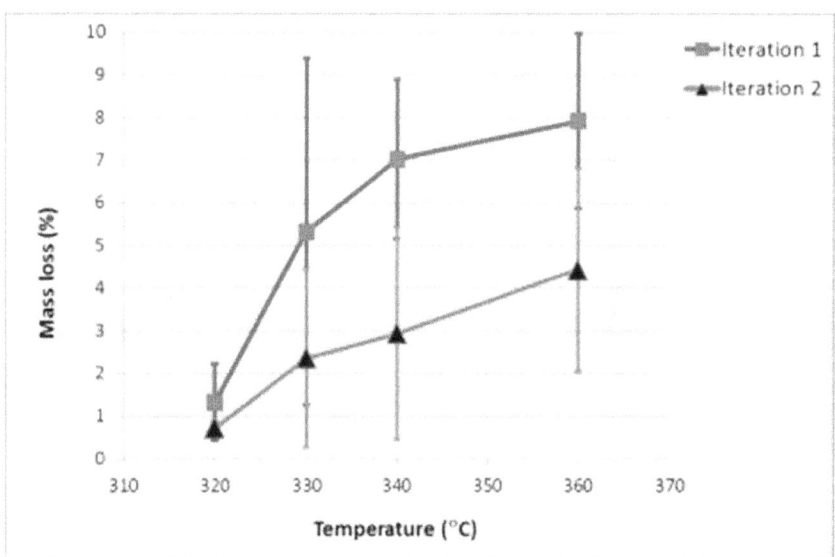

Figure 36. Average mass loss *vs.* set point temperature of the heater in the box test calculated for samples of Iteration 1 and 2. The mass loss is significantly reduced and the onset temperature increased in Iteration 2 as compared to Iteration 1 (uncertainty bars shown here are equal to one StDev).

In Iteration 2, as well as in Iteration 1, the average of the mass loss measured at 340 °C and 360 °C ($ML_{340+360}$) is used as an index of smoldering propensity for a given formulation. In Iteration 2, this average-mass-loss calculation includes also samples that did not show self-sustained smoldering. It can be argued that only self-sustained smoldering samples should be considered, but this does not appear to be feasible because even at higher heater set-point temperatures there would likely be formulations that do not show self-sustained smoldering.[X] The $ML_{340+360}$ values for Iteration 2 formulations are compared in Figure 37.

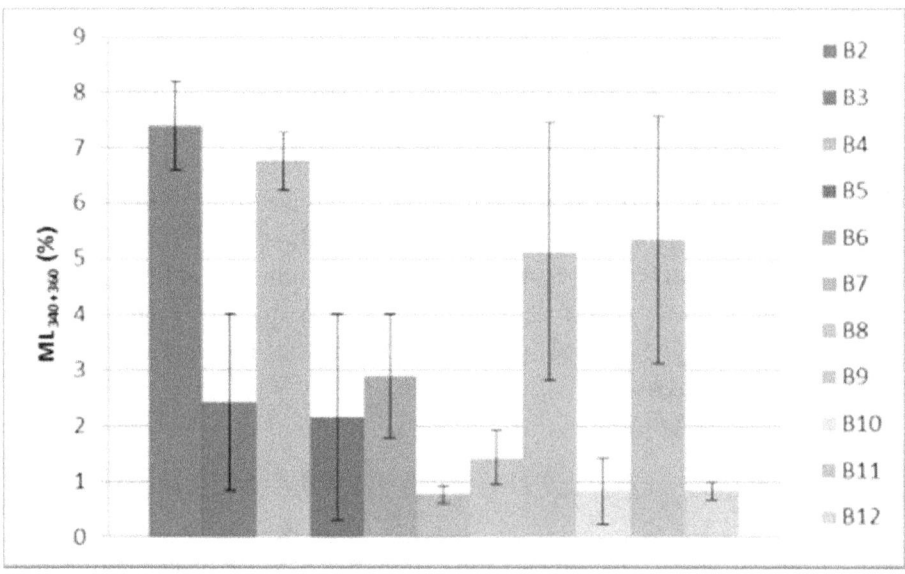

Figure 37. Average mass loss at 340 °C and 360 °C ($ML_{340+360}$) in the box test for the formulations of Table 12 (uncertainty bars equal to one StDev and are calculated over at least 3 replicate measurements).

3.2.3. Correlation between smoldering in the box test and mockup test

The smoldering of formulations from Iteration 2 was also investigated by means of the mockup test; the results are listed in Table 10 (three replicates). A comparison of the smolder behavior measured in the box and the mockup test is shown in Figure 38. Formulations which smoldered moderately in the box test (*i.e.*, $ML_{340+360} < 5$ %) did not smolder significantly in the mockup test and had an air permeability below 45 m·min^{-1} (red data points). As discussed later in Section 3.2.7, an air permeability of 45 m·min^{-1} is a threshold value below which no smoldering is observed in the mockup test. The red line in Figure 38 is a least-squares regression fit to the data points for formulations with an air permeability above this threshold (blue data points). The low value of R^2 ($R^2=0.4$) for the linear fit might be due to an intrinsic poor correlation between the box test and the mockup test, and/or to an effect of in-bun and bun-to bun variability. The linear fit indicates that a mass loss of 30 % in the mockup test (target mass loss for a potential PUF/RM) is roughly equivalent to a mass loss in the box test ($ML_{340+360}$) of 8 %.

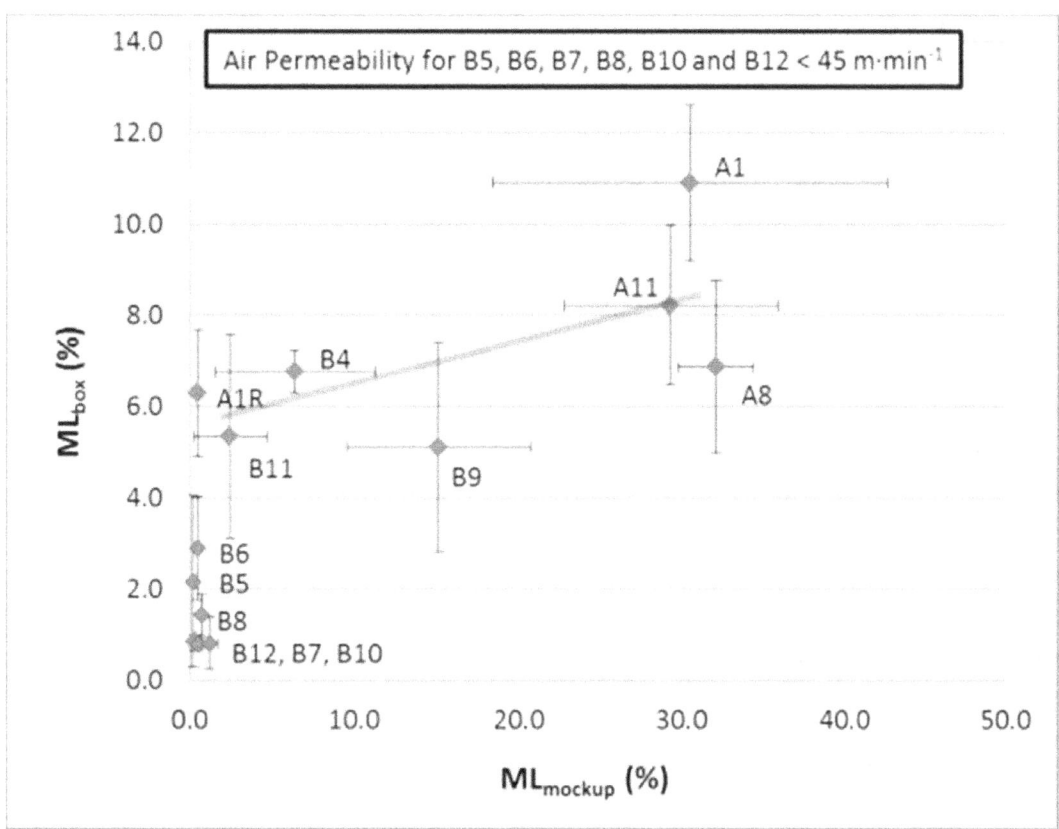

Figure 38. Smoldering measured as mass loss in the box test ($ML_{340+360}$) *vs.* smoldering measured as mass loss in the mockup test (ML_{mockup}) (uncertainty bars shown are equal to one StDev calculated over at least 3 replicates).

The slope of the linear fit of Figure 38 suggests that the mockup test is preferable for foams above the permeability threshold, because in this range it is more sensitive to smoldering variations than the box test. Whereas, below this threshold, the box test is a more useful tool for determining potential smoldering propensity, because smoldering can be boosted by increasing the heater temperature to achieve a significant mass loss even in formulations that are not smoldering in the mockup test.

The relative standard deviation (StDev%) for ML_{mockup} and $ML_{340+360}$ were calculated for three formulations from Iteration 1 (A1, A1-R, A8), nine formulations from Iteration 2 (B4, B5, B6, B7, B8, B9, B10, B11, B12) and three formulations for Iteration 3 (C1, C2, C3). For each formulation, at least three mockup tests and three box tests were run. The average values of StDev% for ML_{mockup} and $ML_{340+360}$ calculated for the samples of Iteration 1 only, Iteration 2 only, Iteration 3 only or all samples from Iteration 1 to 3, are shown in Table 15. In all three iterations, the StDev% in the box test was lower than in the mockup test. The average values of StDev% for all iterations are 34 % for $ML_{340+360}$ and 51 % for ML_{mockup}. The lower reproducibility of the mockup test might be related to variability in the testing materials (*e.g.*, barrier fabric, cotton sheeting, smoking material) other than the foam itself.

The high values of variability observed for these samples (produced in the pilot plant) were largely due to heterogeneity in the samples rather than the specific smoldering test used. As observed in Iteration 4, the relative standard deviations for ML_{mockup} decreased from 51 % in the samples produced in the pilot plant (Iteration 1 to 3 reported in Table 13) to less than 5 % for the samples fabricated in the production line.

Table 13. Average values of relative standard deviation (StDev%) for ML_{mockup} (mockup test) and $ML_{340+360}$ (box test) calculated for the samples of Iteration 1, Iteration 2, Iteration 3 and Iteration 1 to 3.

		$ML_{340+360}$ (Box Test)	ML_{mockup} (Mockup Test)
StDev% for samples from Iteration 1 only	(%)	37	65
StDev% for samples from Iteration 2 only	(%)	42	59
StDev% for samples from Iteration 3 only	(%)	29	64
StDev% for samples from Iteration 1 to 3	(%)	34	51

3.2.4. Effect of the processing parameters on smoldering

Tests conducted with the box and mockup tests per the 2^k3 experimental design (Figure 31), were analyzed to investigate the effect of processing parameters (*i.e.*, tin catalyst, pressure and water level) on smolder propensity.

For better statistical certainty, such a study would require a larger number of replicates and the knowledge of the uncertainty for each of the processing parameters. Practically, this wasn't feasible because the foamer did not provide uncertainties on the processing parameters and the number of tests was limited by time and sample availability. In spite of these limitations, from an engineering and decision-making perspective, this type of analysis was a necessary tool for understanding the critical parameters affecting smoldering and designing the formulations for the following iterations. Similar considerations apply also to Sections 3.2.5-3.2.7.

Data analysis was performed using a software for data mining (FusionPro by S-Matrix) and a cubic regression model to generate 2-D contour plots. The effect of each processing parameter on the response parameter (smoldering performance) was evaluated using a Pareto ranking.[23]

The values of $ML_{340+360}$ and ML_{mockup} were designed as the responses parameters for the box and mockup tests, respectively. The Pareto ranking for the box test data indicated that, in the investigated range, pressure was the parameter with the highest impact on $ML_{340+360}$ and that tin and water had a smaller effect. According to this model, an increase in water level, or a decrease in head pressure and tin content caused an increase in $ML_{340+360}$. A 2-D contour plot for $ML_{340+360}$ vs. tin content and pressure at a low and high water level details these observations in Figure 39 and Figure 40.

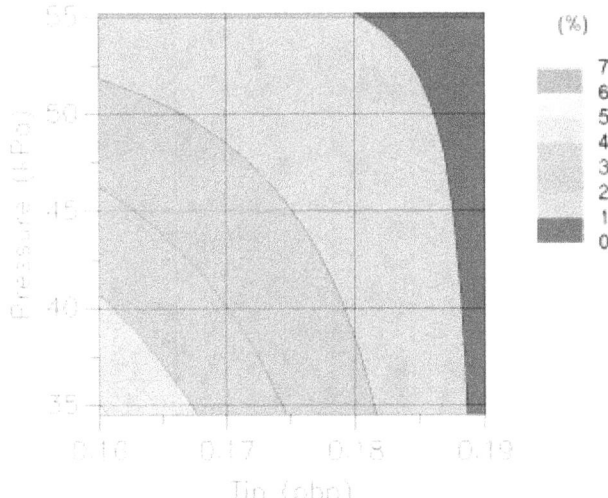

Figure 39. 2-D contour plot for $ML_{340+360}$ vs. tin content and pressure at a low water level (2.70 php).

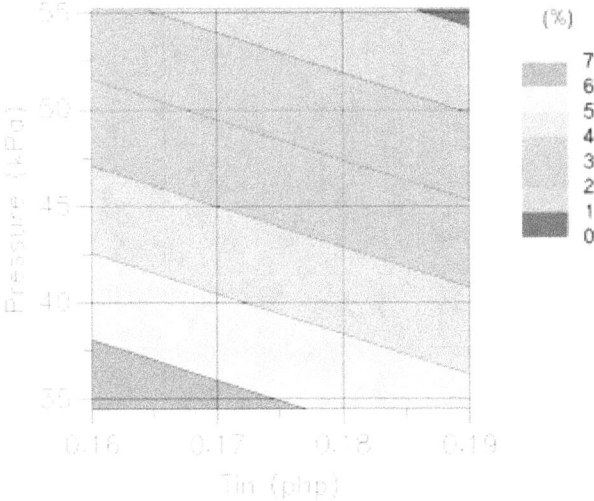

Figure 40. 2-D contour plot for $ML_{340+360}$ vs. tin content and pressure at a high water level (2.95 php).

Similar analyses of the mockup data reveal an analogous effect of tin and pressure on smoldering (*i.e.*, smoldering increases with a decrease in tin catalyst and pressure), and pressure is the parameter with the highest impact on ML_{mockup} (Figure 41 and Figure 42). The water level, however, was had the opposite effect in each test; in the box test, smoldering increased with an increase in water content, whereas, in the mockup test, smoldering decreased with an increase in water content. Supposing that these findings are not biased by data uncertainty, the two tests were affected to a different extent by morphological and processing parameters: smoldering in the mockup test appeared to be more sensitive to a decrease in specific surface area (due to an increase in cell size at high water content), whereas, smoldering in the box test appeared to be more sensitive to an increase in oxygen supply (due to an increase in air permeability at high water content). This is supported by the limiting action on buoyant convection (*i.e.*, oxygen supply) generated by the presence of a box in the box test.

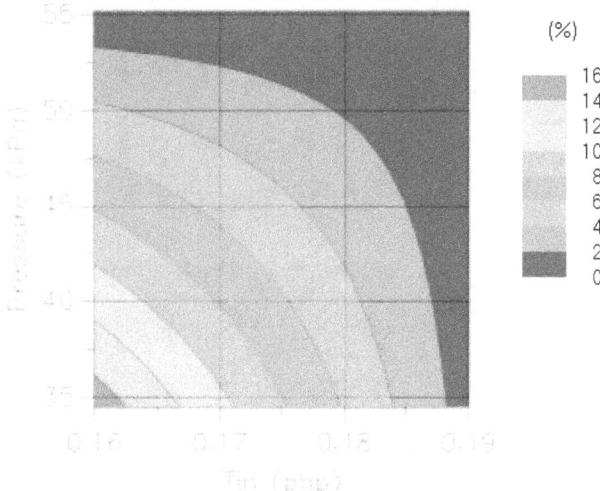

Figure 41. 2-D contour plot for ML_{mockup} *vs.* tin content and pressure at a low water level (2.70 php).

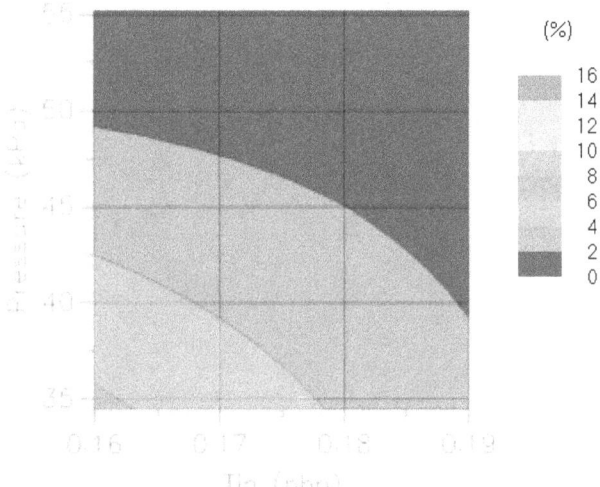

Figure 42. 2-D contour plot for ML_{mockup} vs. tin content and pressure at a high water level (2.95 php).

3.2.5. Effect of the processing parameters on air permeability

The effect of processing parameters on air permeability was investigated with a 2^k3 experimental design (Figure 31). The values of permeability were designed as the response parameter (Table 10) in this analysis.

A cubic regression model was used to generate 2-D contour plots. The Pareto ranking data indicated that, in the investigated range, tin had a major impact on permeability, and that pressure and water had a lower effect. According to this model, 2-D contour plots were generated for air permeability versus tin and pressure at low water level (Figure 43) and high water level (Figure 44).

The data presented in Figure 43 shows that at the low water level of 2.70 php, air permeability was more sensitive to the tin variation rather than the pressure variation. An increase in head pressure or tin content generated a decrease in air permeability.

At a higher water level, 2.95 php, the effect of pressure and tin content on permeability was comparable (Figure 44). An increase in head pressure or an increase in tin content generated a decrease in air permeability.

The comparison between the values of permeability at low water level (Figure 43) and high water level (Figure 44) shows that an increase in water content also promoted an increase in air permeability.

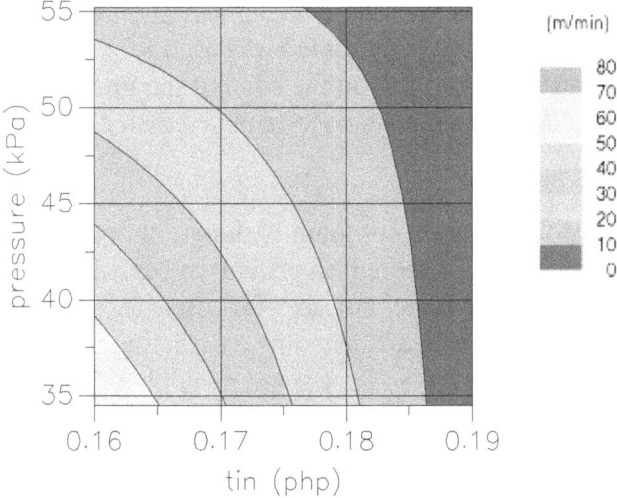

Figure 43. 2-D contour plot for permeability *vs.* tin content and pressure at a low water level (2.70 php). Permeability is slightly more sensitive to the tin variation rather than the pressure variation in the range investigated here and at this water level.

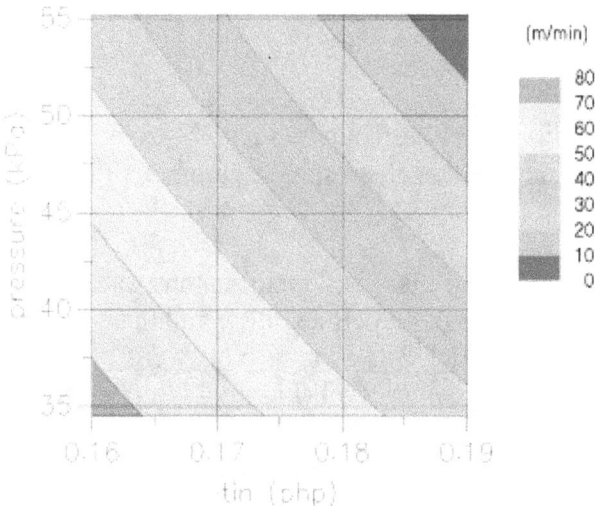

Figure 44. 2-D contour plot for permeability *vs.* tin content and pressure at a high water level (2.95 php). Permeability is equally affected by tin and pressure variations in the range investigated here and at this water level.

3.2.6. Image analysis and cell size determination

The cell size distribution was measured for all formulations used in the 2^k3 experimental design. First, an optical slice with an area of (4.757 x 4.757) mm^2 and a thickness of 197.5 μm was acquired for each formulation by a confocal microscope (Zeiss LSM 510). The images were then processed by an ImageJ plug-in[24] capable to directly segment a gray-level image using a watershed algorithm, also known as Euclidian distance map (EDM).[25] A Gaussian filter blurring with a diameter of 20 pixels was applied to remove noise and prevent over-segmentation before applying the EDM algorithm. Finally, the ImageJ macro "analyze particle" was used for calculating the area of each segmented particle (*i.e.*, cell) that is not on the edge of the image and has a combination of area above 0.02 mm^2 and circularity[XII] above 0.4 (to remove small artifacts and minimize over-segmentation).

As an example, the entire image analysis process for formulation B9 foam is shown in Figure 45. Even though in some cases under or over-segmentation was observed, in general the calculated average values of area per cell (cell area) appeared to be a reliable indicator for cell size. The typical average-cell-area repeatability for this methodology was estimated by calculating the standard deviation over five replicate measurements of Σ in a specific foam bun location. The relative standard deviation for Σ was 10 %. Inhomogeneity in the foam is likely to introduce a more substantial uncertainty. Multiple measurements of Σ would be necessary throughout different buns to account for in-bun and bun-to-bun variability. This was unrealistic at this stage, due to the large number of samples, but it was accomplished in Iteration 4 for those formulations that appeared to be potential SRM/PUF.

[XII] Circularity of a particle is defined as 4π(A/P2), where A and P are the area and the perimeter of the particle, respectively.

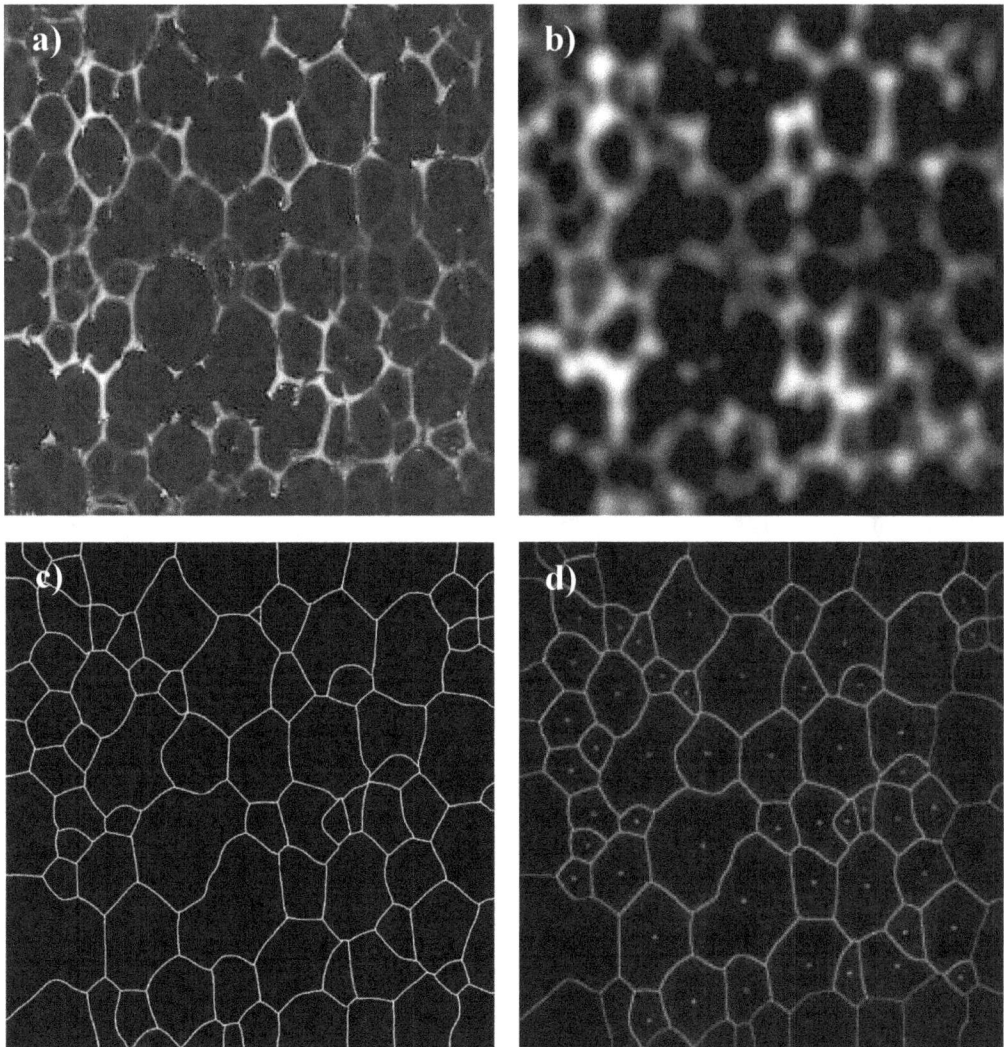

Figure 45. Image analysis and cell area calculation for a B9 foam: a) an optical slice with an area of (4.757 x 4.757) mm^2 and a thickness of 197.5 μm is acquired by a confocal microscope; b) a Gaussian filter is applied to remove noise and prevent over-segmentation; c) segmentation by Euclidian distance map calculation; d) each segmented area that is not on the edge of the image is labeled and measured.

Figure 46 shows the area-per-cell distribution (cell area distribution) and cumulative percentage of the distributions for the same B9 foam. Similar image analyses, cell area distribution and cumulative percentage were obtained for all other formulations (See Appendix 2).

The data are summarized in terms of mean and median values in Figure 47. The mean values of cell area (Σ) are used for ranking the eight formulations in terms of cell size. The cell size ranking is as follows (10 % uncertainty): $\Sigma_{B12} \leq \Sigma_{B9} \approx \Sigma_{B10} \leq \Sigma_{B11} < \Sigma_{B4} \leq \Sigma_{B7} < \Sigma_{B8} \leq \Sigma_{B5}$, ranging from (0.24 ± 0.02) mm^2 for formulation B12 to (0.60 ± 0.06) mm^2 for formulation B5.

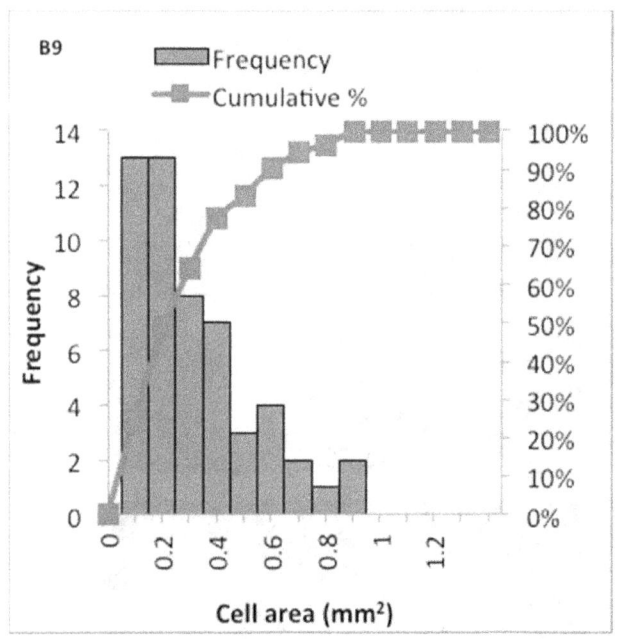

Figure 46. Cell area distribution and cumulative percentage for formulation B9.

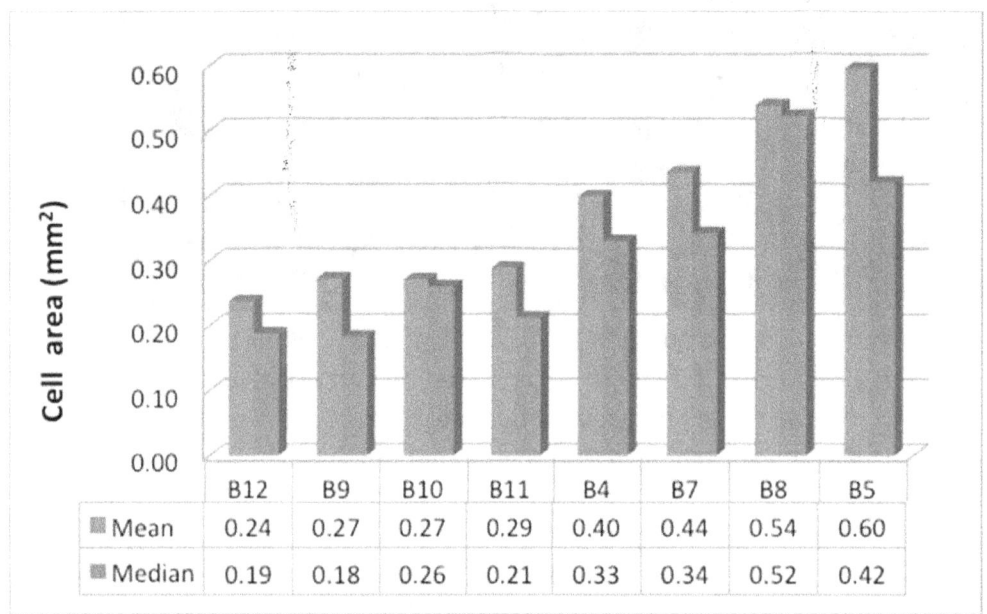

	B12	B9	B10	B11	B4	B7	B8	B5
Mean	0.24	0.27	0.27	0.29	0.40	0.44	0.54	0.60
Median	0.19	0.18	0.26	0.21	0.33	0.34	0.52	0.42

Figure 47. Mean and median values of cell area calculated for all foams of the 2^k3 experimental design.

It is reasonable to deduce that Σ can affect smoldering because as Σ increases, the specific surface (SSA) (area per unit volume of foam) available for oxidation decreases. For example, formulations B5 and B11 have a similar air permeability of about 44 m·min^{-1} but $\Sigma_{B11} < \Sigma_{B5}$, therefore $SSA_{B11} > SSA_{B5}$, demonstrating the expected result that B11 foam is more smolder prone than B5 when tested with both the box and mockup tests (Table 10). The effect of air permeability and SSA will be discussed further in the following sections.

3.2.7. Effect of the processing parameters on cell size

The effect of processing parameters on cell size was investigated in terms of cell area with a 2^k3 experimental design (Figure 31). The mean area per cell, Σ, is specified as the response parameter (Figure 47). A cubic regression model was used to fit the data and generate 2-D contour plots for cell size versus pressure and tin content at low and high water content. The effect of each processing parameter on cell size was evaluated using a Pareto ranking. According to this model, pressure has the highest effect on cell size, followed by tin, whereas the effect of water is marginal. The 2-D contour plot in Figure 48 details the relationship of Σ *vs.* tin content and pressure at a low water level (2.70 php). At this water level, the cell size was dominated by the mixing head pressure; the cell size increased with increase in head pressure.

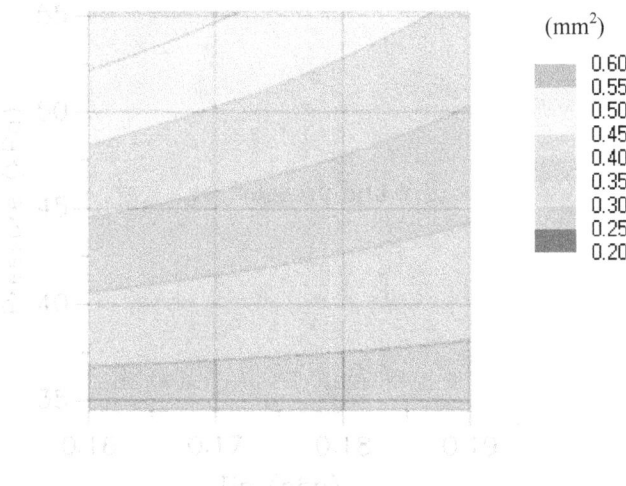

Figure 48. 2-D contour plot for average cell area Σ *vs.* tin content and pressure at a low water level (2.70 php).

Examining the 2-D contour plot in Figure 49 for cell size (Σ) *vs.* tin content and pressure at a high water level (2.95 php) revealed that an increase in head pressure induced an increase in cell size for relatively small values of tin content, but not for high values of tin content. An increase in tin content induced a decrease in cell size.

A comparison of the values of air permeability at a low (Figure 48) and high water level (Figure 49), shows that an increase in water content promoted a small increase in cell size.

These trends can be explained as follows. Foaming in PUF is dominated by the expansion and coalescence of pre-existing air bubbles produced during mixing.[26] The higher the mixing head pressure, the more rapid the expansion of these pre-existing air inclusions at the exit from the mixing chamber due to the increase in differential pressure between the pressure inside the cell and the atmospheric pressure. At a later stage during foaming, water also increases the rate of expansion by generating CO_2 that diffuses inside a cell and increases the differential pressure. The tin catalyst can partially prevent cell expansion by promoting a fast crosslinking and it is reasonable to speculate that this effect is more relevant when the water blowing action is high (*i.e.*, high water level). This might explain why the tin catalyst content has a significant effect only at a high water level.

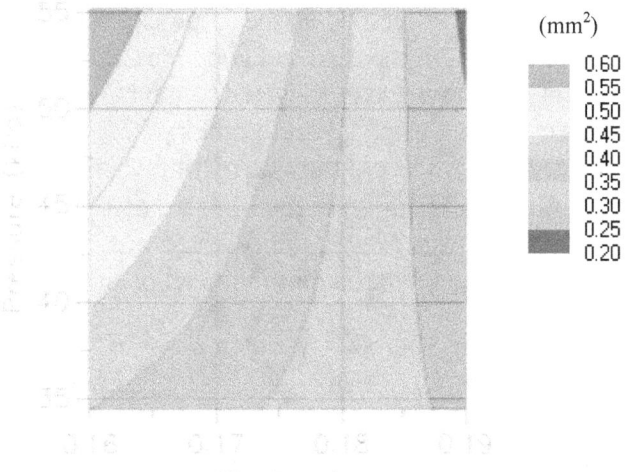

Figure 49. 2-D contour plot for average cell area Σ vs. tin content and pressure at a high water level (2.95 php).

3.2.8. Effect of the air permeability on smoldering

In this section, the effect of air permeability on smoldering is investigated using all available data from Iteration 1 and Iteration 2. These data include the foamer air permeability measurement data and smoldering data from both the box and the mockup tests.

The mass loss in the box test is expressed as the average mass loss measured at a heater-set-point-temperature of 340 °C and 360 °C ($ML_{340+360}$). The values of $ML_{340+360}$ *vs.* air permeability are plotted in Figure 50 (three replicate test per data point). Despite a large data scattering it appeared that $ML_{340+360}$ increases generally linearly with air permeability. An increase in smoldering with air permeability was expected because a higher air permeability induces a larger oxygen supply at the air-foam interface.

Smolder propensity in the mockup test also increases with air permeability (Figure 51) (three replicate test per data point), but in this case, there is a threshold value of air permeability at about 45 m·min⁻¹, below which no smoldering is observed. In this evaluation, a ML_{mockup} greater than 25 % is considered the minimum level for denying a foam as sufficiently smolder prone. Only three formulations resulted in ML_{mockup} above 25 % and have air permeabilities above 70 m·min⁻¹.

The presence of an outlier in Figure 51 ($ML_{mockup} \approx$ 8 % and permeability $\approx$ 75 m·min⁻¹) implies that a permeability above 70 m·min⁻¹ is a necessary but not sufficient specification by itself to achieve a ML_{mockup} of 25 %. In this high range of air permeability, a better morphological description of the foam structure is required to determine other variables that are impacting smolder performance. This investigation is carried out in Iteration 3, which targets formulations with air permeability above 70 m·min⁻¹.

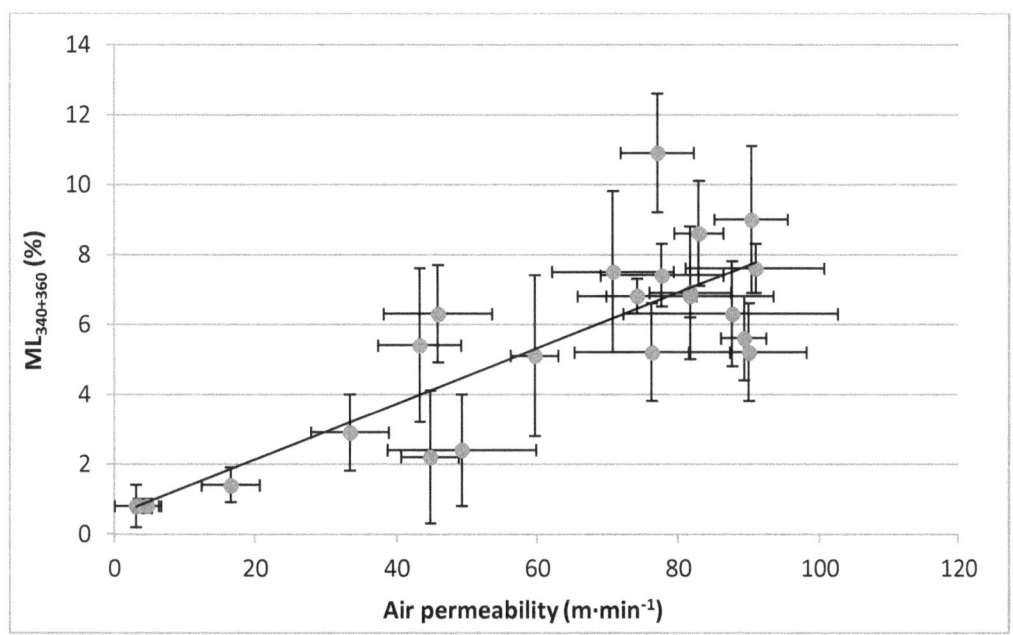

Figure 50. ML$_{340+360}$ (box test) *vs.* air permeability for all available data from Iteration 1 and Iteration 2 (uncertainty bars equal to the StDEv calculated over at least three replicates). The black line is a least-squares regression fit to the data points.

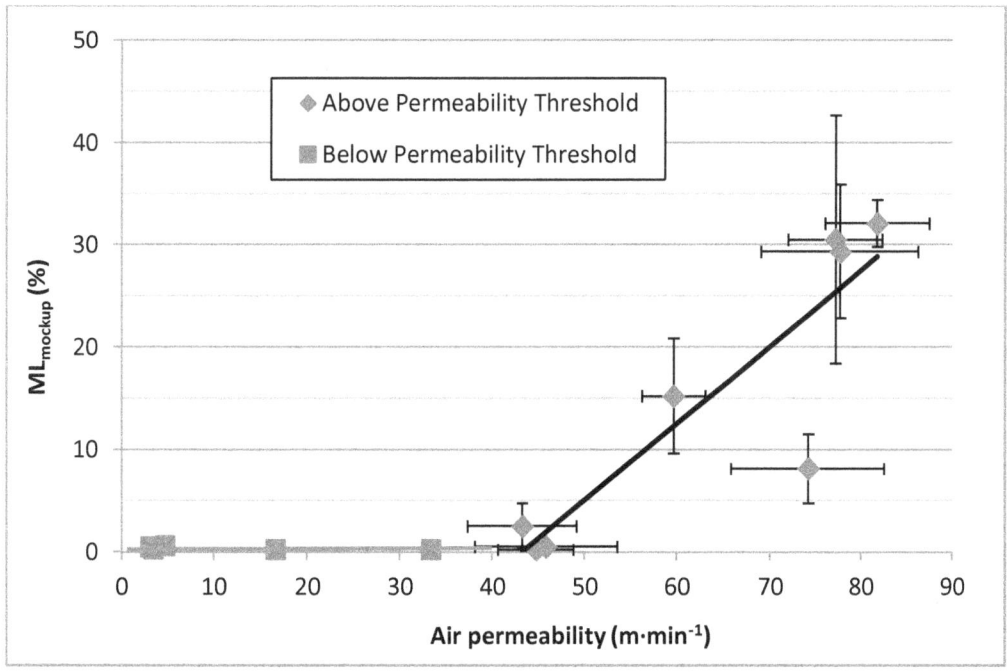

Figure 51. ML$_{mockup}$ (mockup test) *vs.* air permeability for all available data from Iteration 1 and Iteration 2. No smoldering is observed in the mockup test below a threshold value of permeability (uncertainty bars equal to the StDEv calculated over at least three replicates). The red line and black line are least-squares regression fits to the red and blue data points, respectively.

3.2.8 Conclusions of Iteration 2

At the end of Iteration 1, polyol P2 was chosen for continued study. Initial foam buns demonstrated that the variations in manufacturing conditions (specifically, a decrease in temperature and humidity in the pilot plant) caused splitting of the foam. Subsequent foams were made by replacing polyol P2 with polyol P3 and adding a processing aid to prevent foam splitting. The influence of the processing parameters (*i.e.*, water content, catalyst content and head pressure) on smoldering, air permeability and cell size was investigated for a given formulation based on polyol P2 and surfactant S3.

A series of preliminary experiments was performed to identify the processing window, *i.e.*, the maximum range of variability for water content, tin catalyst content and head pressure that ensured successful foaming. A 2^k3 factorial experimental design (two levels for three factors) that examined a high and low level of water, tin catalyst and mixing head pressure was chosen in order to produce flexible foam samples within the specified processing window.

Within the limits of this analysis (unknown uncertainty in the processing parameters, limited number of replicate tests and variability in the foam properties), which are considered acceptable at this stage for engineering and experimental design of the following iterations, the effects of the processing parameters can be summarized as follows:

1. Smoldering behavior:
   - an increase in head pressure or tin content caused a decrease in smoldering;
   - an increase in water content caused an increase in smoldering in the box test and a decrease in the mockup test.
2. Air permeability:
   - an increase in head pressure or tin content induced a decrease in permeability;
   - an increase in water content induced an increase in permeability.
3. Cell size:
   - at a low water level, the mixing head pressure controlled the cell size and the cell size increased with the head pressure; at a high water level, the cell size increased with the head pressure and decreased with the tin catalyst content;
   - an increase in water content induced a marginal increase in cell size.

Smoldering was measured in both the box test and the mockup test. In the box test, smoldering propensity increased with air permeability both in terms of mass loss and onset temperature for sustained smoldering (the higher the air permeability the lower the onset temperature). In the mockup test, no smoldering occurred below an air permeability threshold of about 45 m·min^{-1}. An air permeability value above 70 m·min^{-1} was a required but not necessarily sufficient specification to achieve a mass loss of 25 % in the mockup test. In this high range of air permeability at least one other variable was impacting smoldering. Thus, a better morphological description of the foam structure was required. This investigation was carried out in Iteration 3, targeting formulations with an air permeability above 70 m·min^{-1}.

For foams with an air permeability above 45 m·min^{-1}, it appears to be a poor correlation (R^2 = 0.4) between the box test and the mockup test when a least-squares regression fit is used. The low value of R^2 might be due to an intrinsic poor correlation between the box test and the mockup test, and/or to an effect of in-bun and bun-to bun variability. In a range of permeability above 45 m·min^{-1}, the mockup test appeared to be preferable to the box test due to its higher sensitivity to smoldering variations and a testing configuration that closely mimics a typical smoldering scenario. The box test has proved to be a versatile research tool that can differentiate smoldering propensity also in low permeability foams, provide additional information (onset temperature, temperature profile, power to sustain smoldering), and provide higher repeatability compared to the mockup test.

3.3 Iteration 3: Impact of cell morphology in open-cell PUF

The previous iterations examined the effect of chemistry and processing parameters on the smolder behavior of PUF foams. The third iteration examined the impact of cell morphology on smolder behavior. Data from iteration 2 demonstrated that an air permeability above 70 m·min^{-1} is required for smoldering to occur but an outlier suggests that this is not the only specification required to produce a high smolder prone foam. Iteration 3 targeted PUF that have air permeability above this range to gain a better understanding of the morphological effect of the foam structure on smoldering prone foams. This high value of air permeability can be achieved only in foams with an open cell structure.

Iteration 2 showed that the highest air permeability and smoldering (mockup test) in the 2^k3 experimental design was achieved at a high water level, low tin catalyst level and low pressure level (formulation B4). This is why, in Iteration 3, formulation B4 (P3-S3) was selected as potential SRM/PUF. The three formulations foamed for Iteration 3 are reported in Table 14. Formulation C1 (16 replicate buns) is the replicate of B4. Formulation C2 (14 replicate buns) is identical to C1, but it is foamed at a higher mixing-head pressure because it was found (see Iteration 2) that the head pressure is the main processing parameter controlling smoldering. Formulation C3 (12 replicate buns) is identical to C2 but does not contain the processing aid that was necessary due to the relatively cool temperatures registered during Iteration 2. For each formulation, fourteen replicate buns were foamed in Iteration 3.

Table 14 also reports the average and standard deviation for density and air permeability data for the formulations of Iteration 3 (at least 12 replicates).
In terms of density, the differences between C1 and C2, C1 and C3 are statistically significant (t-test for unpaired data)[27]. Similarly, in terms of permeability, the differences between C1 and C2, and C2 and C3
It appears that an increase in head pressure in C2 and C3 as compared to C1 induces a decrease in the average values of density. Also the addition of the processing aid in C2 might slightly decrease its air permeability as compared to C3, where the processing aid was not used.

Table 14. Composition and processing parameters for the formulations of Iteration 3 (unceratinties are equal to one standard deviation calculated over at least 12 replicates).

Component	Unit	Formulation C1	C2	C3
Polyol P3	php	100	100	100
Processing aid	php	**0.15**	**0.15**	**0**
H_2O	php	2.95	2.95	2.95
Polyether catalyst	php	0.06	0.06	0.06
Amine catalyst	php	0.06	0.06	0.06
Tin catalyst	php	0.16	0.15	0.15
Surfactant S3	php	1	1	1
TDI	php	39.09	39.09	39.09
Mixing head pressure	kPa	**34.5**	**48.3**	**48.3**
		Properties		
Density	$kg \cdot m^{-3}$	30.9±0.9	30.1±0.8	29.9±0.6
Permeability	$m \cdot min^{-1}$	78.4±4.6	70.9±10.7	78.5±6.9

3.3.1. Smoldering vs. air permeability

All of the Iteration 3 foams were tested in the box test at multiple set point temperatures, the data are reported in Table 15. The average smoldering behavior measured in the box test, $ML_{340+360}$, for all formulations of Iteration 3 is (5.9 ± 3.5) % and it is comparable to the average $ML_{340+360}$ measured in iteration 1 (7.3 ± 1.5) %. Sustained smoldering was never observed for C2, even at a heater temperature of 360 °C and the value of $ML_{340+360}$ was significantly lower as compared to C1 and C3 (Figure 52).

Table 15. Mass loss measured in the box test for formulation of Iteration 3 at a heater-set-point-temperature of 320 °C, 330 °C, 340 °C, and 360 °C.

	MASS LOSS (%)									
	320 °C		330 °C		340 °C		360 °C		340 °C+360 °C[†]	
	Single Test	AVG[‡] (Dev)	Single Test	AVG[‡] (Dev)	Single Test	AVG[‡] (Dev)	Single Test	AVG[‡] (Dev)	AVG	StDev
C1	1.0[§]	1.0	8.1	6.7	10.1	9.0	9.1	9.1	9.1	1.1
	-	-	5.3	1.4	7.9	1.1	-	-		
C2	0.9[§]	0.9	1.0[§]	0.8	1.5[§]	1.3	3.9[§,*]	3.9	2.2	1.5
	-	-	0.5[§]	0.3	1.1[§]	0.2	-	-		
C3	0.9[§]	0.9	0.8[§]	1.0	6.2	6.3	7.0	7.0	6.5	0.4
	-	-	1.1[§]	0.2	6.5	0.2	-	-		

[†]Average mass loss calculated at 340 °C and 360 °C. [‡]Average (AVG) and deviation (Dev) calculated at a given target temperature. [§]Non-smoldering sample. *This sample showed an anomalous behavior: it did not show self-sustained smoldering even if $ML_{340+360}$ was above 2 %.

The values of $ML_{340+360}$ *vs.* air permeability are plotted for all available data from Iteration 1, Iteration 2 and Iteration 3 in Figure 53. The formulations of Iteration 3 had similar air permeability but dramatically different smoldering performance. This indicates that at least for foams with high permeability there is another parameter controlling smoldering. A similar conclusion can be drawn by looking at the smoldering data in the mockup test (Table 16).

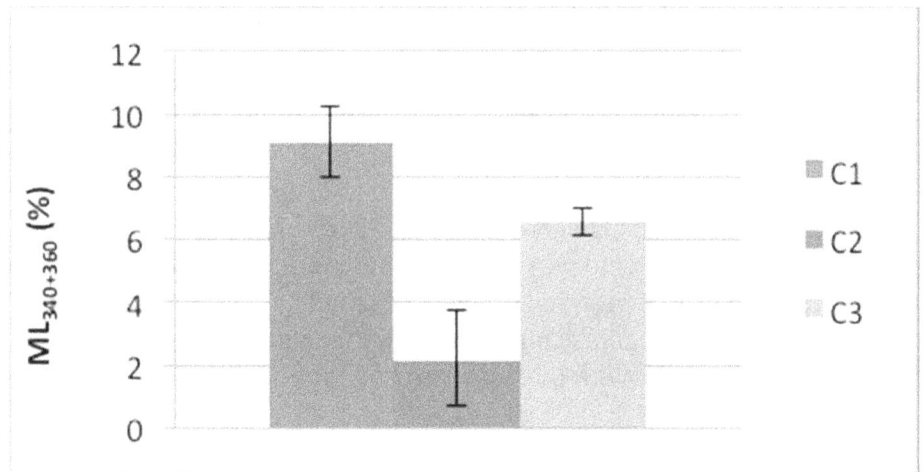

Figure 52. Average mass loss at 340 °C and 360 °C ($ML_{340+360}$) in the box test for the formulations of Iteration 3 (uncertainty bars equal to one standard deviation and are calculated over at least 3 replicate measurements).

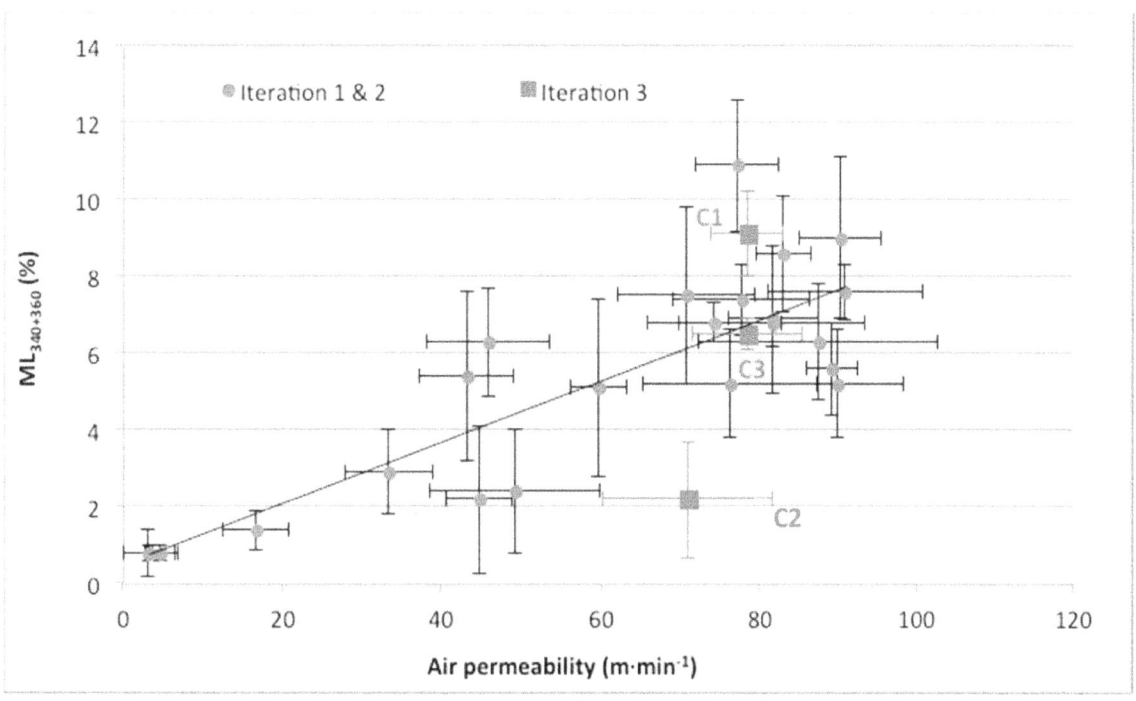

Figure 53. $ML_{340+360}$ (box test) vs. air permeability for all available data from Iteration 1, Iteration 2 (blue data points), and Iteration 3 (red data points). The black line is a least-squares regression fit to the blue data points (uncertainty bars equal to one standard deviation calculated over at least three replicates).

Table 16. Mass loss measured in the mockup test for formulations of Iteration 2.

		C1	C2	C3
ML_{mockup}	(%)	21.5±10.6†	10.0±2.5†	1.2±0.3†

†Uncertainty is equal to one standard deviation calculated over at least three replicates.

The values of ML_{mockup} as a function of air permeability are plotted for all available data from Iteration 1, Iteration 2, and Iteration 3 in Figure 54. Formulation C1 showed the highest mass loss in the box test ($ML_{340+360}$) and mockup test (ML_{mockup}), however, C3 had a higher mass loss than C2 in the box test and, vice versa, C2 had a higher mass loss than C3 in the mockup test. The differences in mass loss measured between C2 and C3 are significant. These results indicate that the box test cannot be used as a robust predictor for the mass loss in the mockup test (additional testing is required to avoid possible bias due to foam variability). The presence of a wood box in the box test might dramatically decrease buoyant convection and oxygen supply, and thus affect smoldering, especially in foams with high air permeability.

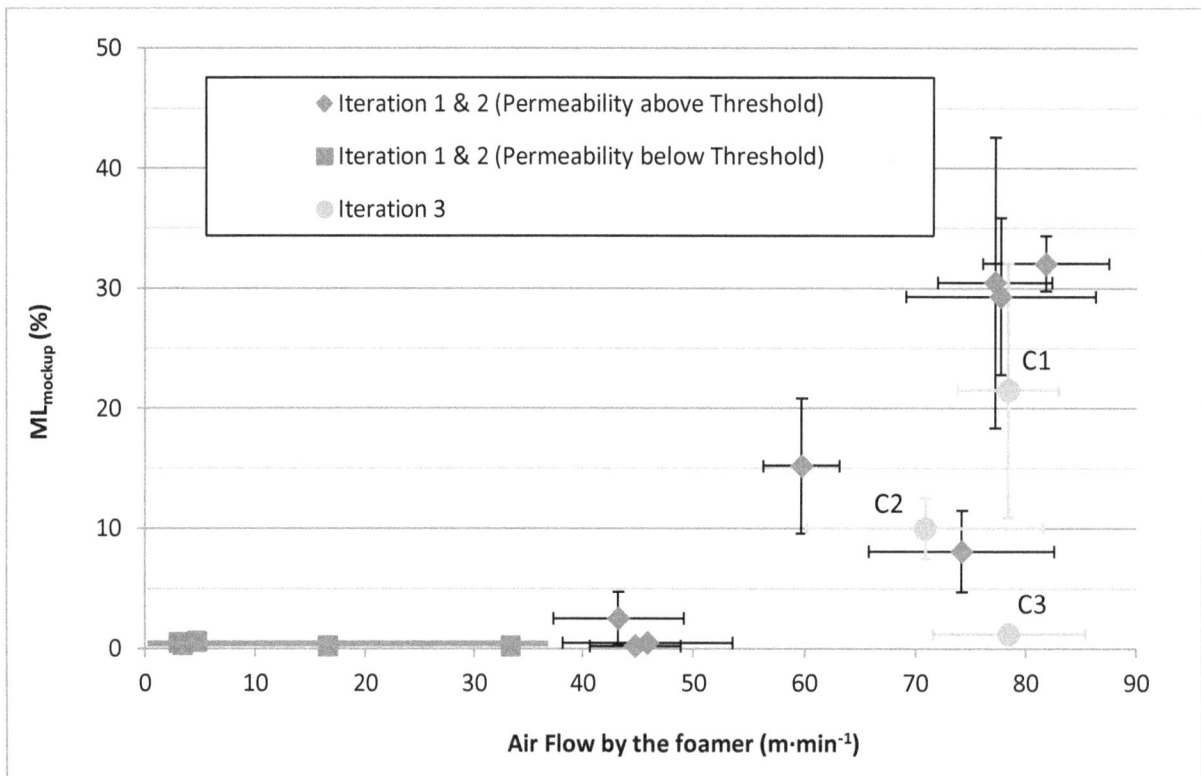

Figure 54. ML_{mockup} (mockup test) *vs.* air permeability for all available data from Iteration 1, Iteration 2 and Iteration 3. The formulations of Iteration 3 had similar air permeability but dramatically different smoldering. This is indicating that at least for foams with high permeability there is another parameter controlling smoldering (uncertainty bars equal to one standard deviation calculated over at least three replicates). The red line is a least-squares regression fit to the red data points.

The values of ML_{mockup} shown in Figure 54 are each an average value measured for a given formulation. Formulation C1 was selected to investigate the effect of bun-to-bun variability on smoldering behavior because it had the highest average ML_{mockup} of Iteration 3. Sixteen different foams were prepared for formulation C1 (see Appendix 1) and characterized. Figure 55 shows the ML_{mockup} for formulation C1 measured on several buns with different air permeability. In this high range of permeability, smoldering appeared to increase when air permeability decreased. This was an unexpected result, contrary to the common belief that smoldering always increases with air permeability.[28]

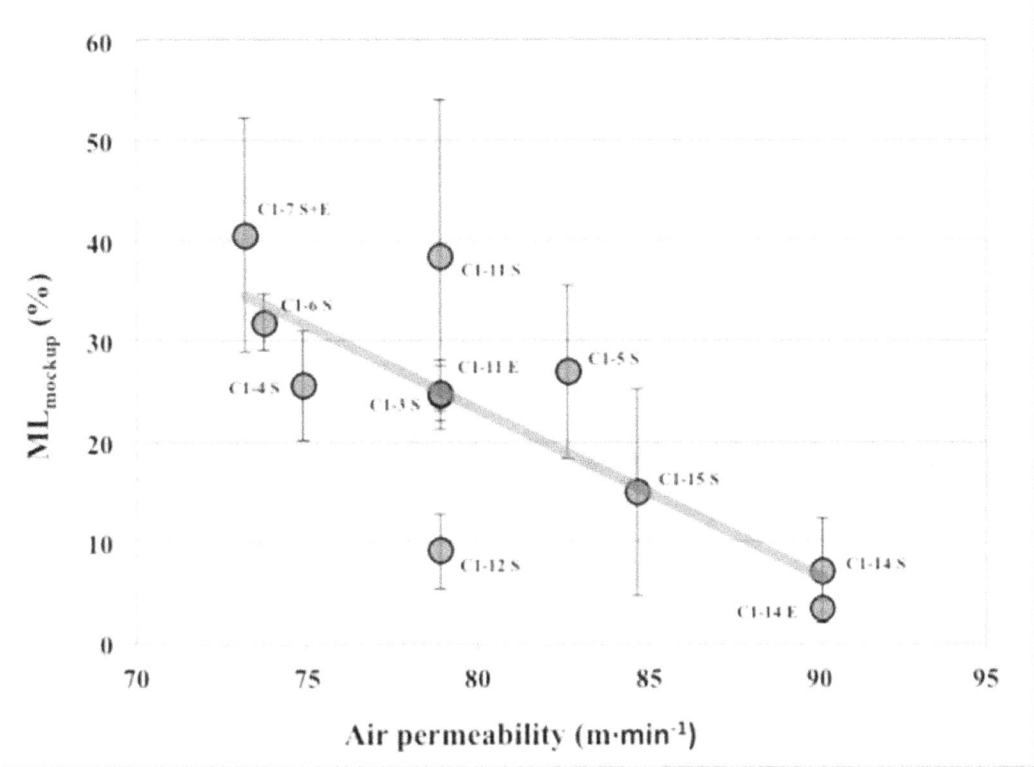

Figure 55. ML_{mockup} (mockup test) as a function of air permeability for formulation C1 measured for several foam buns. In this high range of permeability smoldering increases when air permeability decreases. The red line is a least-squares regression fit to the data points (uncertainty bars are equal to one standard deviation calculated over at least three replicates).

Reviewing the data in Figure 54 and Figure 55, infers that:
- to achieve a ML_{mockup} greater than 2 %, an air permeability (Φ) above a threshold value of about 45 $m \cdot min^{-1}$ ($\Phi_{threshold}$) is a necessary but not sufficient specification,
- for $\Phi > \Phi_{threshold}$, there is no clear correlation between air permeability and ML_{mockup}
- at high air permeability ($\Phi > 70\ m \cdot min^{-1}$), ML_{mockup} increases with a decrease in Φ for formulation C1.

To understand this phenomenon a morphological study of the foams from Iteration 3 was conducted in the following sections.

3.3.2. Effect of Specific Surface Area on Smoldering

The data shown above emphasize the need for a better morphological description of the foam structure. In this section, the effect of specific surface area on smoldering in the mockup test is investigated. The specific surface area (SSA) on smoldering is defined here as surface per unit volume and is expressed in inverse meters. The total surface of PUF was calculated by density and BET measurements (see Section 2.5).[29]

The calculated values of SSA for all formulations of Iteration 3 and some formulations of Iteration 1 and 2 are reported in Table 17 together with air permeability (Φ) and smoldering data (ML$_{mockup}$). Each value of SSA was measured in a specific foam bun location. Two different foam samples, collected from the same bun and location, were used for replicate measurements in C1-3S and C1-5S. The maximum absolute deviation for SSA was about 15 % from the mean value. Since the typical uncertainty for BET measurements is below 1 %, significant heterogeneities existed in terms of SSA even in nearly the same foam bun location.[30]

Table 17. Values of specific surface area (StDEv and AVG calculated over at least three replicates).

	Air permeability* Φ (m·min^{-1})	Specific surface area SSA (m^{-1})		ML$_{Mockup}$ (%)	
		Avg	StDev	Avg	StDev
C1-3S	79	3839	19	24.6	3.4
C1-3S	79	4898	18	24.3	0.2
C1-4S	75	3693	15	25.5	5.4
C1-5S	83	4994	58	26.9	9
C1-5S	83	3316	3	26.9	8.6
C1-6S	74	4132	16	31.8	2.8
C1-11S	79	3880	16	38.4	15.5
C1-11E	79	3264	16	24.8	2.7
C1-12S	79	3105	19	9.1	3.7
C1-14S	90	2928	15	7.1	5.2
C1-14E	90	3320	6	3.4	0.6
C1-15S	85	3496	28	14.9	10.2
C2-6S	75	2677	15	11.8	13.2
C2-7S	83	3316	24	8.2	6.1
C3-9S	90	2783	9	1	0.1
C3-8S	85	3339	9	1.6	1.9
C3-12S	75	3144	18	1.1	0.5
A1-1S	75	4543	81	30.5	12.1
B5-2E	42	3844	28	0.2	0.1
B11-2M	46	5720	188	1.6	0.9
B12-3S	1	5173	270	0.3	0.3
B4-4E	70	4742	31	6.4	4.9

*These values of air permeability were measured by the foamer in the middle section of the bun and do not take into account possible in-bun variations.

The relative standard deviation calculated for SSA over all C1 samples of Table 17 (12 samples), affected by both bun-to-bun and in-bun variability, was about 20 %. This value of standard deviation was also assumed for formulations A1, C2 and C3 since the available data for SSA were not sufficient to calculate a standard deviation. Similarly, for ML_{mockup} the standard deviation calculated for C1, 50 % was taken as a typical standard deviation for the other formulations.

The value of ML_{mockup} is plotted as a function of SSA in Figure 56. There are three different datasets:

1. data for a specific foam bun and location with high air permeability ($\Phi > 70$ m·min^{-1});
2. data for a specific foam bun and location with low air permeability ($\Phi \leq 70$ m·min^{-1});
3. average data for a specific formulation with high air permeability ($\Phi > 70$ m·min^{-1}).

Dataset 1 shows that ML_{mockup} increased with SSA even though there was a large scatter in the data, likely due to variability in foam properties. The averaged dataset 3 showed a similar trend with less scatter. Dataset 2 did not show any correlation with ML_{mockup}.

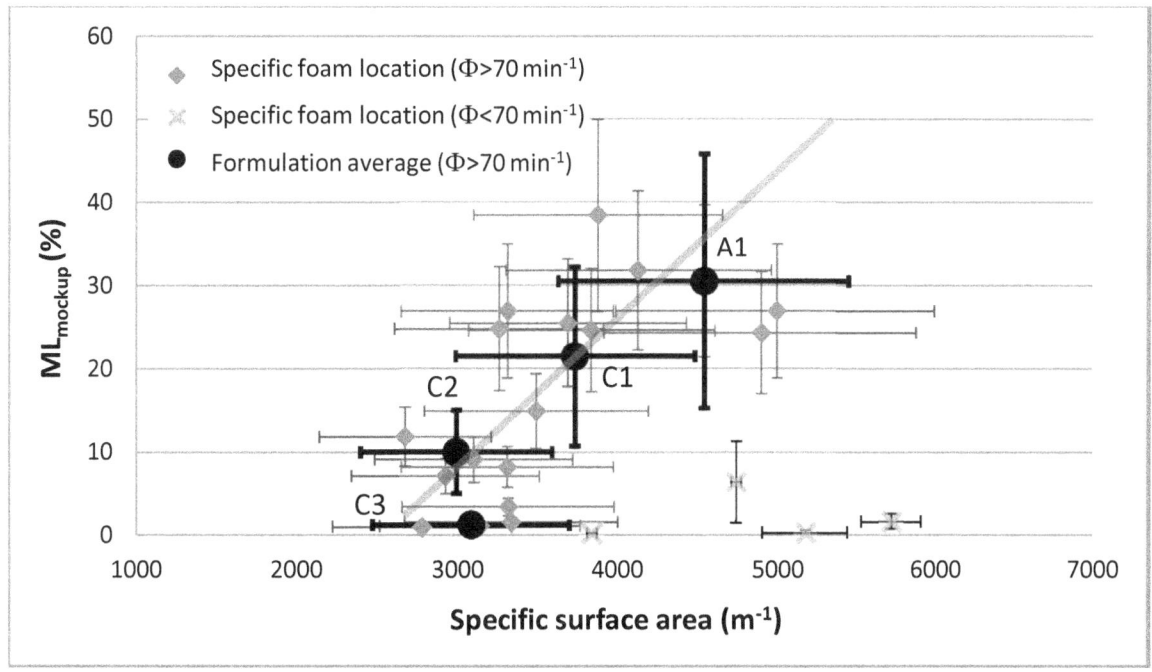

Figure 56. ML_{Mockup} vs. specific surface area. Uncertainty bars are equal to one StDev and calculated from at least 3 replicate measurements for datasets 1 (blue dots) and 2 (green dots); for dataset 3 (black dots), as previously described, the standard deviation were assumed equal to 20 % and 30 % of the mean value of SSA and ML_{mockup}, respectively. The red line is a least-squares regression fit to the black data points.

The highest value of specific surface area was observed for B11-2M (SSA = 5720 m^{-1}). However, for sample B11-2M, as well as samples B5-2E, B12-3S and B4-4E (dataset 2) the air permeability was relatively low, *i.e.*, $\Phi \leq 70$ m·min^{-1}. For these foams, the fraction of closed cell membranes was likely higher than the one for foams with $\Phi > 70$ m·min^{-1}, and a large fraction of SSA was generated by residual membranes in the cells. In fact, a BET measurement accounts for the surface generated by the struts and the cell membranes. In particular, a BET measurement accounts also for the surface of the cell membrane in a closed cell foam because krypton gas will diffuse over time into a closed cell.[30] Similarly, oxygen can diffuse through a cell membrane and sustain smoldering, but the oxygen supply rate will be lower in a closed cell as compared to an open cell, where oxygen transport is dominated by convection rather than diffusion. Furthermore, residual closed membranes in open cells can dramatically reduce oxygen supply by suppressing air convection. This is likely why formulations with high SSA but low Φ (*e.g.*, B11-2M) showed limited smoldering.

The importance of Φ is confirmed in Figure 57 showing the values of ML_{mockup} as a function of specific surface area for formulation C1 measured on several buns. In this high range of permeability, the number of residual cell membranes was minimal and smoldering increased as the specific surface area increased. Noticeably, data of ML_{mockup} as a function of air permeability for the same Iteration 3 foam buns (Figure 55), showed that smoldering increased as air permeability decreased.

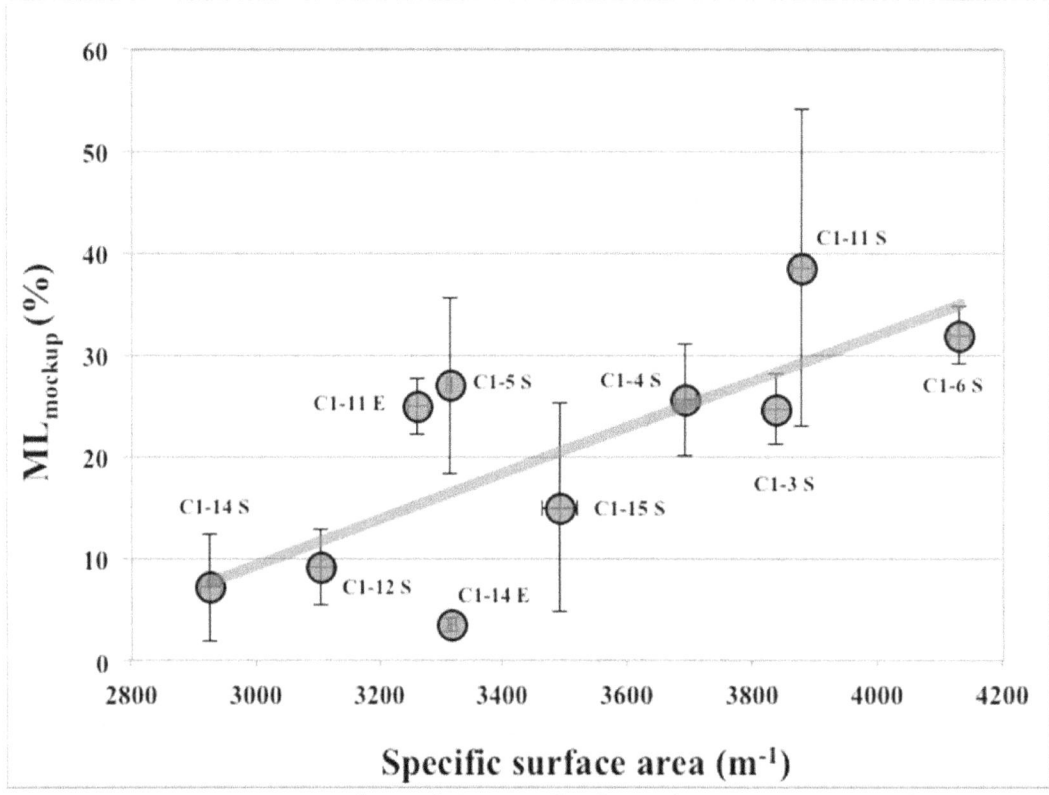

Figure 57. ML_{mockup} *vs.* specific surface area for formulation C1 measured on several buns. The red line is a least-squares regression fit to the data points. In this high range of permeability, smoldering increased when the specific surface area increased (uncertainty bars equal to one standard deviation calculated over at least three replicates). The data of ML_{mockup} *vs.* air permeability for the same foam buns are shown in Figure 55.

In general, an increase in air permeability is promoted by an increase in fraction of open windows and an increase in cell size. In completely open-cell foams (*i.e.*, no residual closed membrane), a further increase in permeability can be achieved by increasing the cell size, for example, by increasing the mixing head pressure. This effect is dominant in a high permeability range where most of the cells are open.

This principle is illustrated in the schematic drawing of
Figure 58. Either increasing the fraction of open membranes (case *a*) or increasing the cell size (case *b*) induces an increase in air permeability and a decrease in SSA. However, smoldering increases for case *a*, due to an increase in oxygen supply, and decreases for case *b*, due to a reduction in SSA. For a given formulation, the morphology of the foam that maximizes smoldering is characterized by a fine and largely open cell structure with a high value of air permeability (*i.e.*, $\Phi > 70$ m·min^{-1}). A high value of Φ is necessary to promote high oxygen supply through convective movements. In this high range of permeability a decrease in cell size promotes smoldering by increasing the amount of air/foam interface available for oxidation.

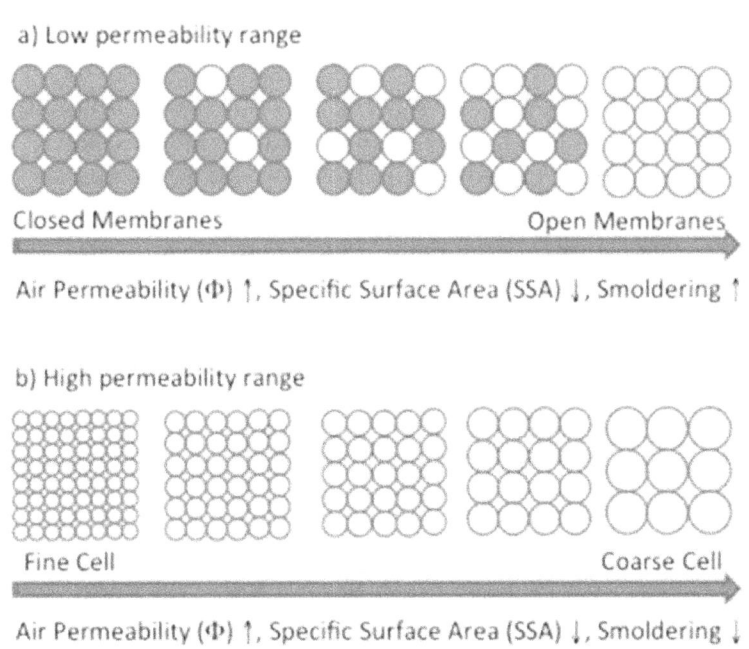

Figure 58. Schematic drawing illustrating two possible mechanisms promoting an increase in air permeability (Φ): a) increase in fraction of open membranes; b) increase in cell size. Both mechanisms induce also a decrease in SSA but only a) promotes an increase in smoldering due to an increase in oxygen supply.

In general, in a typical PUF, where both open and closed cells coexist, there is no clear correlation between SSA and smoldering. In practice, this implies that SSA is a good morphological descriptor for smoldering propensity only in fully open PUF, like reticulated PUF[XIII].

3.3.3. Image analysis and cell size determination

As discussed in the previous section, the specific surface area appears to be a good morphological descriptor for smoldering propensity only in fully open-cell PUF. In addition, BET measurements (necessary for SSA calculations) are extremely time-consuming and it is difficult to perform more than 2-3 measurements per day. Due to the bun-to-bun and in-bun variability, multiple measurements are necessary for a statistically sound approach. In practice, this is unrealistic.

In this section, cell size is considered as alternative morphological descriptor for smoldering propensity. Noticeably, cell size is independent on the fraction of closed cells. The average cell structure is measured in terms of cross-sectional cell area by means of image analysis; then, the effect of cell size on smoldering is discussed. The assumption, here, is that average cell area is related to the SSA as illustrated in
Figure 58 (case b), and a detailed study of cell size distribution is not required.

The average cell area of C1, C2, and C3 formulations was measured with an approach similar to the one described in Iteration 2. However, in this case, the scanned area for each image was larger (142 mm^2 instead than 50.4 mm^2) and for each foam bun, three confocal images were acquired from orthogonal planes to account for possible anisotropy in the foam:
- image 1 (top view) is a confocal image acquired from the top surface of the bun after removing the foam skin;
- image 2 (side view 1) and image 3 (side view 2) are confocal images acquired from the two planes which are orthogonal to the top surface plane and to each other.

Even though in some cases an under/over-segmentation was observed, in general, the calculated average values of cell area appeared to be a reliable indicator for cell size. Inhomogeneity in the foam likely introduced a substantial variability in smoldering behavior.

A complete cell-size analysis (including scan time and image analysis) required about an hour, rather than the several hours required for BET and density measurements.

The buns selected for this study were C1-6M[XIV] and C1-14M (investigating the variation of cell structure in the same formulation), plus C2-3M and C3-4M (investigating the effect of mixing pressure and processing-aid). The confocal images from top view and side views with the relative image analysis and cell-area-distribution histograms for the four buns are reported in Appendix 2.

[XIII] A reticulated foam is a foam that is post-processed to remove all residual membranes by chemical etching or thermal treatment.
[XIV] M indicates that the foam sample was collected from the middle section of the bun, see Section *Formulation, bun and location identification.*

The typical average-cell-area uncertainty for this methodology was estimated by calculating the standard deviation over five replicate measurements of Σ in a specific foam bun location (formulation C1, sample size of approximately 125 cm^3). The relative standard deviation for Σ was 4 % in a specific foam location but, for a formulation in general, it can be significantly higher due to in-bun and bun-to-bun variability.

The confocal micrograph images in Appendix 2 indicate that in this range of air permeability all foams have similar open-cell structures, but there are obvious variations in cell size.

The following parameters are calculated for all foams and summarized in Table 18:
- Σ_{TOP}: average cell area measured from top view;
- Σ_{Side1}: average cell area measured from side 1 view;
- Σ_{Side2}: average cell area measured from side 2 view;
- Σ_{Side}: average of Σ_{Side1} and Σ_{Side2};
- Σ: average of Σ_{Side1}, Σ_{Side2} and Σ_{TOP}.

Table 18. Average cell area calculated from each view point.

	Average cell area (mm^2)				
	Σ_{TOP}[†]	Σ_{Side1}[‡]	Σ_{Side2}[‡]	Σ_{Side}[*]	Σ[§]
C1-6M	0.21	0.26	0.25	0.26	0.24
C1-14M	0.42	0.42	0.48	0.45	0.44
C2-3M	0.34	0.31	0.43	0.37	0.36
C3-4M	0.40	0.47	0.37	0.42	0.41

[†]Average cell area measured from top view.

[‡]Average cell area measured from side view 1 or side view 2.

[*]Average of Σ_{Side1} and Σ_{Side2}.

[§]Average of Σ_{TOP}, Σ_{Side1} and Σ_{Side2}.

For a given bun, generally the average value of cell area is larger in the side view (Σ_{Side}) than in the top view (Σ_{TOP}) due to foam anisotropy. The average cell area Σ, calculated as the average of Σ_{Side1}, Σ_{Side2} and Σ_{TOP}, was used for cell size quantification.

The cell size ranking is as follows: $\Sigma_{C1-6M} < \Sigma_{C2-3M} < \Sigma_{C3-4M} < \Sigma_{C1-14M}$. The cell area per cell varied from a minimum value of 0.24 mm^2 for C1-6M to a maximum of 0.44 mm^2 for C1-14M. The minimum and maximum values of Σ were observed for the same formulation (C1), which highlights the poor repeatability and large bun-to-bun variability in these formulations produced in the pilot plant.

The effect of cell size on mass loss in the mockup test was evaluated with data (three replicates) from foam buns C1-6 and C1-14 of formulations C1 (subject to in-bun variability), and all data for formulations C2 and C3 in Figure 73 (subject to bun-to-bun and in-bun variability).[XV] The

---

[XV] An average value of ML$_{mockup}$ for formulations C2 and C3 is reported because less than three replicate mockup tests were available for the specific foam buns C2-3 and C3-4.

standard deviation shown here for Σ is assumed to be equal to 4 % and it does not consider in-bun and bun-to-bun variability.

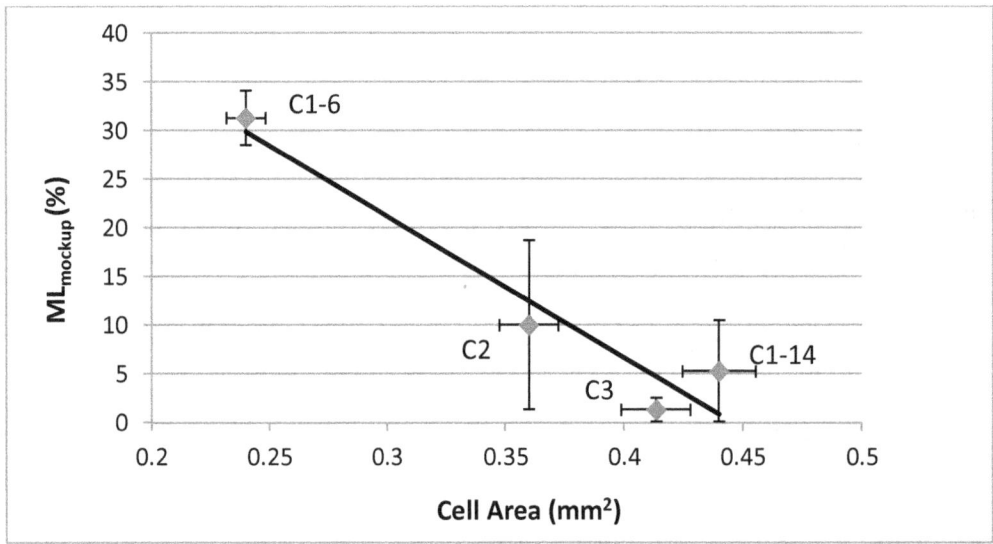

Figure 59. Effect of cell size (Σ) on smoldering in the mockup test (uncertainty bars equal to one StDev). The black line is a least-squares regression fit to the data points.

There is a clear effect of Σ on ML_{mockup}: the larger the cell size, the lower the smoldering mass loss. The linear fit of Figure 59 was used to estimate the value of Σ generating a ML_{mockup} of 30 % ($\Sigma \approx 0.239$ mm^2) as a potential standard reference material foam (SRM/PUF). If a variation of ± 15 % in smoldering mass loss is acceptable for a SRM/PUF then the corresponding acceptable variation in Σ, calculated by using the linear fit of Figure 59, is approximately ± 13 % (*i.e.*, $\Sigma = 0.239 \pm 0.031$).

At this stage, the aforementioned correlation between cell area and ML_{mockup} requires further validation due to unknown in-bun and bun-to-bun variability. Given the large variability in the foams produced in the pilot plant, it was decided to postpone the validation of these data in Iteration 4, while foams with much more consistent properties were prepared on a production line.

3.3.4. Open-flame Tests (bare foam)

The open-flame resistance of bare foams was measured according to the open-flame resistance test described in the CPSC proposed standard.[15] Three formulations from Iteration 3 (C1, C2, C3) and one formulation from Iteration 2 (B5) were tested. The tests qualify a foam's stand-alone flaming performance in a bare foam configuration and evaluate a fire-barrier fabric with a cover fabric.

A test conducted with formulation C1 demonstrated melt dripping of the foam. Figure 60 shows a test with formulation C1 a few seconds before extinguishing the sample. Although flaming liquid material was dripping to the underlying tray, there was no pool fire with sustained flaming that could have boosted the heat release rate though a feed-back effect (Figure 61).[31] A similar behavior was observed in all samples showing sustained combustion after removing the ignition source. Two formulations (B5 and C2) did not show sustained flaming after ignition source removal. When ignition was not observed after the first 5 s of flame impingement, a second attempt was carried out by applying the open-flame for another 5 s in a new location, 13 cm (5 in) away from the original ignition point. For both formulations (B5 and C2), the second attempt also failed. Figure 61 shows a mockup assembly for formulation C2 after attempting ignition twice. A plot of the mass loss as a function of time for formulations C1 and C3 are shown in Figure 62 and Figure 63, respectively. Three tests per formulation are reported. Specific data on each test are available in Appendix 3.

Values for sustained flaming (ignition), final mass loss percentage 120 s after ignition source removal (ML_{120s}) and foam properties (density and air permeability) are reported in Table 19. The ML_{120s} values for C1 and C2 are identical and slightly below the target value of 20 % specified in CPSC proposed standard, section 1634.24 (Figure 81 and Figure 82, respectively). Three tests per formulation are reported. Specific data on each test are available in Appendix 3.

Figure 60. A test with formulation C1 few seconds before extinguishing the sample. At this stage flaming liquid material is dripping to the underlying tray but there is no pool fire with sustained flaming that can boost the heat release rate through a feed-back effect.

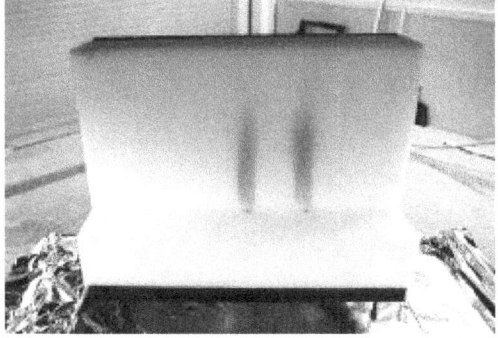

Figure 61. Ignition was not observed after the first 5 s of flame impingement with formulations B5 and C2; a second attempt was carried out by applying the open-flame for other 5 s in a new location 5 in. apart from the original ignition point. Here, it is shown a mockup assembly for formulation C2 after attempting ignition twice.

Table 19. Summary of the open-flame tests on bare foams. The values of densities and permeability shown here are the averages of four values for B5, 16 values for C1, 14 values for C2 and 12 values for C3. The uncertainty is equal to one standard deviation.

Formulation	Buns tested	Ignition	ML_{120s} (%)	Density $(kg \cdot m^{-3})$	Air Permeability $(m \cdot min^{-1})$
B5	B5-3	NO	0	30.4±0.3	44.8±4.1
C1	C1-8 & C1-9	YES	16.7±3.5	30.9±0.9	78.4±4.6
C2	C2-9	NO	0	30.1±0.8	70.9±10.7
C3	C3-3 & C3-1	YES	16.7±3.2	29.9±0.6	78.5±6.9

Only two of four formulations ignited when exposed to the open-flame even though the composition of the foams was substantially the same. However, the structure and morphology (*e.g.,* cell size, fraction of open/close cells, etc.) were different, indicating that these characteristics played an important role not only on smoldering but also on ignition. In particular, it is reasonable to conclude that in this fire scenario where convective heat transfer is dominant, ignition is more likely to occur in formulations with a high value of air permeability (Table 19).

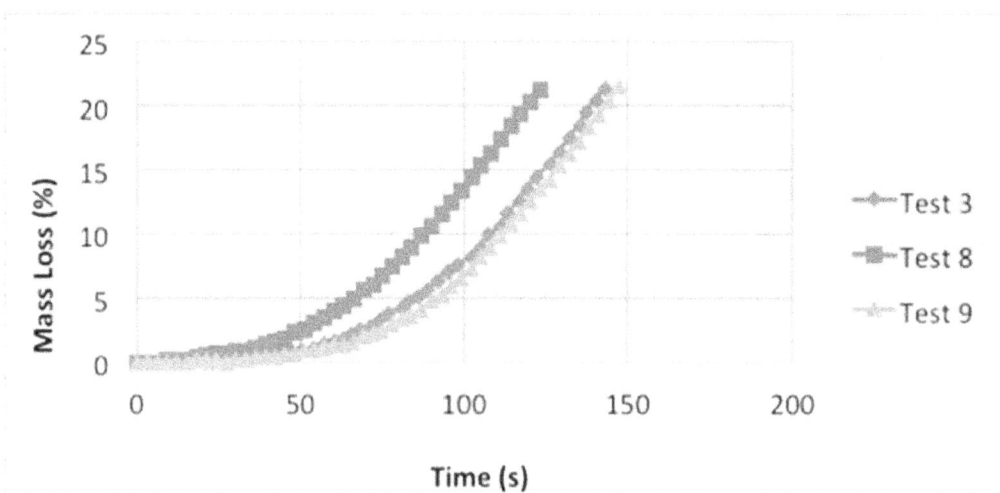

Figure 62. Mass loss *vs.* time measured for three mockups prepared with formulation C1.

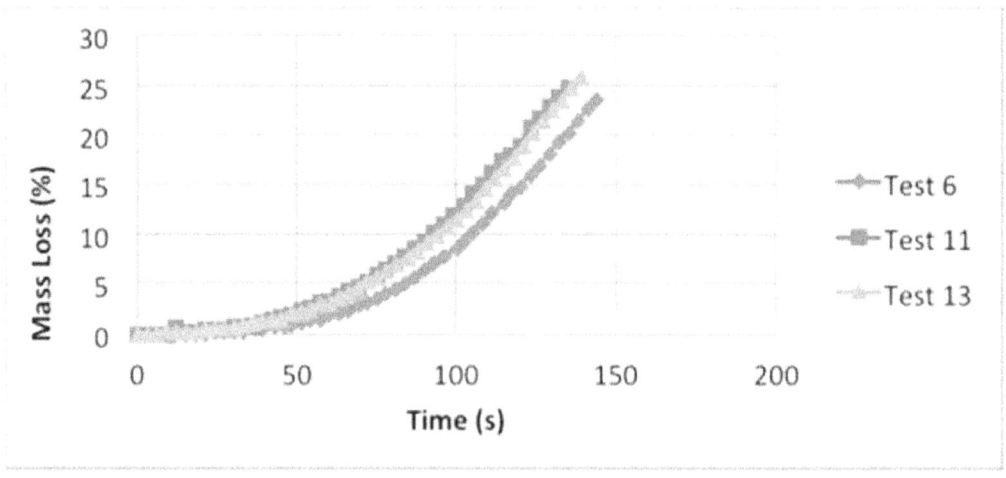

Figure 63. Mass loss *vs.* time measured for three mockups prepared with formulation C3.

3.3.5. Open-flame Tests with barrier material

Three samples from formulation C1 were tested in the Open-flame Test with barrier material (see Tests 7, 10 and 12 in Appendix 3). The final result of the ignition can be seen in Figure 64 and Figure 65, which show the entire mockup and only the foam of the same mockup assembly, respectively. The foam blocks appeared to be partially charred on the surface but they retained their shape and original mass. As seen in Figure 66, mass loss was about 1 % when the pilot flame was removed. According to the CPSC proposed standard, a barrier material fails the test if the final mass loss reaches 20 % of the original mass of the foam (fire barrier and fabric are not counted in the mass loss determination). The average final mass loss calculated from the 3 tests was (6.3 ± 1.2) % and was due only to mass loss in the fire barrier material, polyester fabric and cover fabric. The lack of a significant mass loss in the foam blocks demonstrates the effectiveness of the barrier material. This effectiveness is clearly shown in Figure 67 which compares the residue of mockups tested with the fire barrier material (no foam mass loss after 45 min) and the residue of the mockups tested without the fire barrier material ((16.7 ± 3.5) % foam mass loss) after two min.

Figure 64. Photo of the residue of the mockup after the open-flame test with barrier material.

Figure 65. Photo of the residue of the mockup at the end of the test after removing the barrier material, the polyester fabric and the cover fabric.

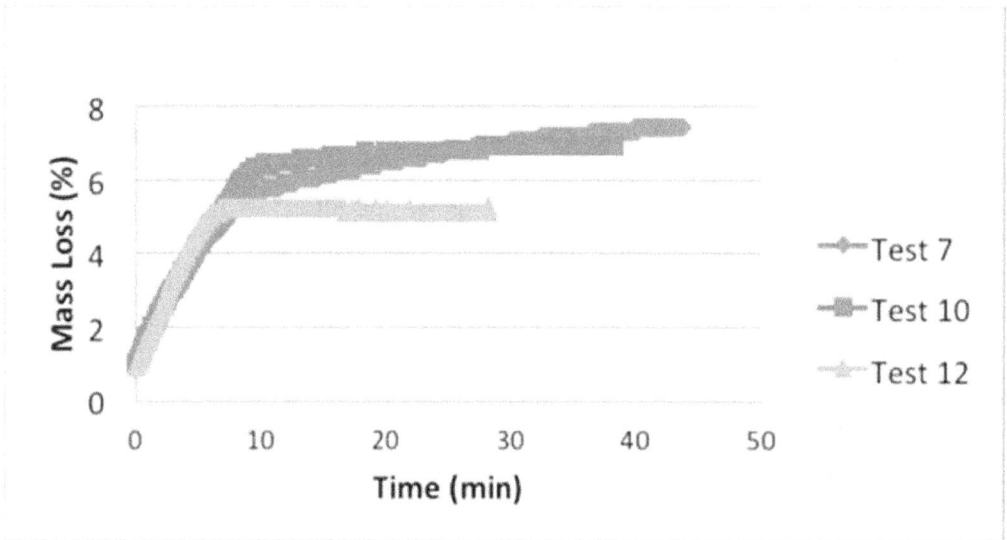

Figure 66. Total mass loss in barrier test as a funciton of time measured for three mockups prepared with formulation C1.

Figure 67. Photo of the comparison between residues of mockups (formulation C1) tested with barrier material (on the left) and without barrier material (on the right).

3.3.6. Conclusions of Iteration 3

The data in Iteration 2 led to the conclusion that a variable other than air permeability was impacting smoldering and that a better morphological descriptor of the foam structure was required. This investigation was carried out in Iteration 3 on formulations with high air permeability (above 70 m·min⁻¹). Both mockup and box tests indicated that high air permeability was a necessary but not sufficient condition to achieve intense smoldering. For these foams, the smoldering data measured by the box test could not be used to predict smoldering behavior in the mockup test. The smoldering scenario is substantially different in these two tests. It is presumed that the box in the box test significantly reduces the oxygen supply to the smoldering front, especially in high permeability foams.

An increase in permeability caused a decrease in smoldering propensity for values of air permeability above 70 m·min^{-1}. This was a new finding that required careful characterization of the cell morphology. NIST developed a confocal microscopy technique based on self-fluorescence of PUF that allowed comparisons of optical slices of foams with identical optical thickness. This is a key requirement for a proper comparison of two dimensional cell sizes, where the apparent cell size is a function of the optical thickness of the slice. Image analysis was used for measuring the distribution of area cell and the average value of the surface area per cell was used as a morphological descriptor for cell size.

The specific surface area data measured by BET analysis and density showed that smoldering is a function of specific surface area and that, at least for PUF with high air permeability (above 70 m·min^{-1}), the oxygen supply to the smoldering front is not a limiting factor. As an alternative to specific surface area, cell size can also be used as a morphological indicator for smoldering propensity. Cell size measurements are relatively fast and not affected by the fraction of closed cells. Mockup mass loss appeared to increase with decreased cell size in foams with high air permeability.

PUF with reduced smoldering propensity can be prepared by controlling the head pressure during the foaming process; an increase in pressure increases cell size and a decrease in surface area, causing a decrease in smoldering propensity. Adjusting the head pressure during foaming enables tuning of the specific surface area, thus reducing the smoldering propensity in PUF without any required change in formulation.

Open-flame tests showed that, in a fire scenario where convective heat transfer is dominant, ignition is more likely to occur in formulations with a high value of air permeability. This might also imply that is possible to increase ignition resistance of PUF by decreasing air permeability (*i.e.*, promoting a closed cell structure). Open-flame tests, showed also that the use of a barrier fabric was extremely effective and prevented any mass loss in the foam during the test.

At this stage, the feasibility of a standard reference material with a well characterized and reproducible smoldering is linked to the ability of manufacture to product a PUF with a well-defined and reproducible cell size.

3.4. Iteration 4: From Pilot Plant to Production Line

NIST is collecting, analyzing, and interpreting characterization and smoldering data of PUFs produced from several buns from the foamer's production/manufacturing line. The results (a complete report will be provided in a separate document) indicate it is possible to commercially produce PUFs with reproducible smoldering. Preliminary data show that an approximate 18% smoldering mass loss PUF can be produced by targeting a C1 formulation with a permeability of (71 ± 8) m·min^{-1}, cell area of (0.31 ± 0.01) mm^2, and density of (27.4 ± 0.2) kg·m^{-3} (uncertainties equal to one standard deviation).

4. Summary

This 30 month project was broken into four iterations. In Iteration 1, NIST determined the smoldering attributes of PUF were not significantly impacted by the raw materials (*i.e.,* polyol and surfactant) used to produce PUF in the foamer's pilot plant.

The foamer attempted to reproduce a promising formulation from Iteration 1, but due to changes in climate, an additional processing aid was required to produce foam without macroscopic defects in the pilot plant.

In Iteration 2, NIST determined the processing parameters (*i.e.,* tin catalyst content, water content, and equipment head pressure) strongly influenced the smoldering attributes of the PUF. NIST also determined how changes in these parameters impacted the PUF physical attributes, correlated these physical characteristics to smoldering and determined target values for these physical characteristics, which yield high smoldering. General trends were observed, such as decreased head pressure or tin catalyst results in high permeability and high permeability results in high smoldering. However, there were exceptions where PUF with similar permeability had significantly different smoldering performance. NIST measured cell size using confocal microscopy and showed smaller average cell sizes resulted in higher smoldering.

NIST also developed and assessed a Box Smoldering test that was expected to enable measuring the PUF smoldering performance without the use of other materials (*e.g.,* cover fabric). This is a robust test for measuring smoldering propensity but, in its current configuration, is not a good predictor for CPSC mockup test, and NIST is currently considering modifications.

In Iteration 3, NIST determined the cell morphology was significantly impacted by the processing conditions and developed/utilized techniques (*e.g.,* BET and confocal microscopy) to measure changes in the cell morphology. The data indicate the degree of PUF smoldering mass loss can be adjusted by targeting specific values of permeability, cell size, and foam density. Based on the data from this research project, two examples using a C1 type formulation are as follows.

An approximate 30 % smoldering mass loss PUF can be produced by targeting a
- permeability > 75 m·min^{-1},
- cell size $= (0.24 \pm 0.02)$ mm^2, and
- density $= (30.4 \pm 1.6)$ kg·m^{-3}.

An approximate 18 % smoldering mass loss PUF can be produced by targeting a
4. permeability $= (71 \pm 8)$ m·min^{-1},
5. cell area $= (0.31 \pm 0.01)$ mm^2, and
6. density $= (27.4 \pm 0.2)$ kg·m^{-3}.

Additional effort will be required to identify a manufacturer with quality controls sufficient to produce a long-term supply of foam consistent with CPSC's proposed test standard.

5. References

[1] Ahrens, M., *U.S. home structure fires*. 2011, National Fire Protection Association: Quincy, MA. p. 118.

[2] Ahrens, M., *U.S. fires in selected occupancies*. 2006, National Fire Protection Association: Quincy, MA. p. 318.

[3] National Fire Protection Association, Coalition for Fire-Safe Cigarettes, http://www.nfpa.org/categoryList.asp?categoryID=2254 or www.firesafecigarettes.org

[4] Ihrig, A.M., et al., *Factors involved in the ignition of cellulosic upholstery fabrics by cigarettes*. Journal of Fire Sciences, 1986. 4(4): p. 237-260.

[5] Dwyer, R.W., Fournier, L.G., Lewis, L.S., Furin, D., Ihrig, A.M., Smith, S., Hudson, W.Z., Honeycutt, R.H., and Bunch, J.E., *The effects of upholstery fabric properties on fabric ignitabilities by smoldering cigarettes*. Journal of Fire Sciences, 1994. **12**(3): p. 268-283.

[6] Singh, H. and Jain, A.K., *Ignition, combustion, toxicity, and fire retardancy of polyurethane foams: a comprehensive review*. Journal of Applied Polymer Science, 2009. **111**(2): p. 1115-1143.

[7] Weil, E.D. and Levchik, S.V., *Commercial flame retardancy of polyurethanes*. Journal of Fire Sciences, 2004. **22**(3): p. 183-210.

[8] Ohlemiller, T.J., *Modeling of smoldering combustion propagation*. Progress in Energy and Combustion Science, 1985. **11**(4): p. 277-310.

[9] Ohlemiller, T.J., *Smoldering combustion*. in the SFPE Handbook of Fire Protection Engineering (3rd edition; Eds.: DiNenno, P.J., Drysdale, D., Beyler, C.L., and Walton, W.D.). 2002, National Fire Protection Association: Massachusetts. p. 2.200–2.210.

[10] Rogers, F.E., Ohlemiller, T.J., Kurtz, A., and Summerfield, M., *Studies of the smoldering combustion of flexible polyurethane cushioning materials*. Journal of Fire and Flammability, 1978. **9**: p. 5-13.

[11] Dodd, A.B., Lautenberger, C., and Fernandez-Pello, C., *Computational modeling of smolder combustion and spontaneous transition to flaming*. Combustion and Flame, 2012. **159**(1): p. 448-461.

[12] Ortiz-Molina, M.G., Toong, T-Y., Moussa, N.A., and Tesoro, G.C., *Smoldering combustion of flexible polyurethane foams and its transition to flaming or extinguishment*. Symposium (International) on Combustion, 1979. **17**(1): p. 1191-1200.

[13] Rein, G., Lautenberger, C., Fernandez-Pello, A.C., Torero, J.L., and Urban, D.L., *Application of genetic algorithms and thermogravimetry to determine the kinetics of polyurethane foam in smoldering combustion*. Combustion and Flame, 2006. **146**(1-2): p. 95-108.

[14] Torero, J.L. and Fernandez-Pello, A.C., *Forward smolder of polyurethane foam in a forced air flow*. Combustion and Flame, 1996. **106**(1-2): p. 89-109.

[15] *Proposed standard for the flammability of upholstered furniture*, 73 FR 11702 (March 4, 2008).

[16] Ohlemiller, T.J., On the Criteria for Smoldering Ignition in the CFR 1632 Cigarette Test for Mattresses, in NIST Technical Note 1601. 2008, National Institute of Standards and Technology: Gaithersburg. p. 18.

[17] http://www.nist.gov/customcf/get_pdf.cfm?pub_id=902075

[18] British Standard BS 5852:2006 – *Methods of test for assessment of the ignitability of upholstered seating by shouldering and flaming ignition sources (composite mock-ups)*.

[19] Brunauer, S., Emmett, P.H., and Teller, E., *Adsorption of gases in multimolecular layers.* Journal of the American Chemical Society, 1938. **60**(2): p. 309-319.

[20] Gustafsson, S.E., *Transient plane source techniques for thermal conductivity and thermal diffusivity measurements of solid materials.* Rev Sci Instrum 1991. 62(3): p. 797–804.

[21] Housel, T., *Flexible polyurethane foam.* Chapter 5 in Handbook of Polymer Foams (Ed.: Eaves, D.), Rapra Technology Limited, 2004, p. 85-122.

[22] O'Connor, J. M. and Russell, D., *Understanding polyurethanes.* Smithers Scientific Services Inc., training seminar, Cleveland, Ohio, April 8-10, 2008.

[23] Belegundu, A. D. Et al. *Multi-objective Optimization of Laminated Ceramic Composites Using Genetic Algorithms.* Proceedings of the 5th AIAA/NASA/USAF/ISSMO Symposium on Multidiscipinary Analysis and Optimization, Washington D.C.,1994, p. 1015-1022.

[24] http://rsbweb.nih.gov/ij/

[25] http://bigwww.epfl.ch/sage/soft/watershed/

[26] Klempner, D. and Sendijarevic, V., *Fundamentals of foam formulation.* in Handbook of Polymeric Foams and Foam Technology (2nd edition; Eds.: Klempner, D. and Sendijarevic, V.). Hanser Gardner Publications, Cincinnati, 2004, p. 5-14.

[27] http://www.itl.nist.gov/div898/handbook/eda/section3/eda353.htm

[28] Rein, G., *Smouldering combustion phenomena in science and technology.* International Review of Chemical Engineering, 2009, p. 3-18.

[29] Brunauer, S., Emmett, P.H., and Teller, E., *Adsorption of gases in multimolecular layers.* Journal of the American Chemical Society, **60**(2), p. 309-319 (1938).

[30] Pendleton, P. and Badalyan A., *Adsorption,* 2005, 11 p. 61-66.

[31] Ohlemiller T.J. and Shields J.R., *Aspects of the fire behavior of thermoplastic materials.* NIST Technical Note 1493, 2008.

www.ingramcontent.com/pod-product-compliance
Lightning Source LLC
Chambersburg PA
CBHW081831170526
45167CB00007B/2791